HIGHER EDUCATION
AND SDG2

HIGHER EDUCATION AND THE SUSTAINABLE DEVELOPMENT GOALS

Series Editor

Wendy Purcell
Emeritus Professor and University President Emerita, and Academic Research Scholar with Harvard University

About the Series

Higher Education and the Sustainable Development Goals is a series of 17 books that address each of the SDGs, in turn, specifically through the lens of higher education. Adopting a solutions-based approach, each book focuses on how higher education is advancing delivery of sustainable development and the United Nations global goals.

Forthcoming Volumes

Higher Education and SDG11: Sustainable Cities and Communities edited by Julio Lumbreras and Jaime Moreno-Serna

Higher Education and SDG4: Quality Education edited by Tawana Kupe

Higher Education and SDG16: Peace, Justice and Strong Institutions edited by Sarah E. Mendelson

Higher Education and SDG10: Reduced Inequalities edited by Priya Grover, Nidhi Phutela, Pragya Singh

HIGHER EDUCATION AND SDG2

Zero Hunger

EDITED BY

KAREN CRIPPS
Oxford Brookes University, UK

AND

PARIYARATH SANGEETHA THONDRE
Oxford Brookes University, UK

emerald
PUBLISHING

United Kingdom – North America – Japan – India
Malaysia – China

Emerald Publishing Limited
Emerald Publishing, Floor 5, Northspring, 21-23 Wellington Street,
Leeds LS1 4DL.

First edition 2024

Editorial matter and selection © 2024 Karen Cripps and
Pariyarath Sangeetha Thondre.
Individual chapters © 2024 The authors.
Published under exclusive licence by Emerald Publishing Limited.

Reprints and permissions service
Contact: www.copyright.com

British Library Cataloguing in Publication Data
A catalogue record for this book is available from the British Library

ISBN: 978-1-83608-461-7 (Print)
ISBN: 978-1-83608-458-7 (Online)
ISBN: 978-1-83608-460-0 (Epub)

INVESTOR IN PEOPLE

CONTENTS

List of Figures and Tables   ix

Series Editor Preface   xi

Acknowledgements   xv

1   An Introduction to SDG2 Zero Hunger
    Karen Cripps and Pariyarath Sangeetha Thondre   1

Part 1: The Global South Research-Based Policy and Community Perspectives

2   Urban Resilience from Agriculture: A Case Study of
    Ho Chi Minh City in Vietnam
    Quyen Vu Thi and Meri Juntti   19

3   Consequences of Heavy Rain on Productivity of
    Sugar Companies in Peru: A Case Study Towards
    Achieving SDG2 Zero Hunger
    Juan Diego Zamudio Padilla and Constanza Flores Henríquez   43

4   Food Security, Nutrition, and Sustainable Agriculture Nexus:
    The Role of Higher Education in Attainment of Zero Hunger
    in Zimbabwe
    Prosper Chopera, Tonderayi Mathew Matsungo,
    Sandra Bhatasara, Viren Ranawana, Alberto Fiore,
    Faith Manditsera, and Lesley Macheka   65

5   Contribution of Lilongwe University of Agriculture and
    Natural Resources to the Attainment of SDG2 in Malawi
    Agnes Mbachi Mwangwela, Vincent Mlotha,
    Alexander Archippus Kalimbira, William Kasapila,
    Jessica Kampanje Phiri, Samuel Mwango, and
    Samson Pilanazo Katengeza   87

    Case Study: Diverse Characteristics of Daily Dietary
    Practices of Chinese Urban Residents: The Case
    of Guangzhou
    Lin Jiahui and Zeng Guojun   110

          Part 2: The Global North: Teaching and Learning,
                Governance and Community Outreach

6   The Potential of Campus Food Gardens to Achieve
    Student Food Literacy and Security in Australia
    Sophia Lin, Cathy Sherry, Tema Milstein,
    Seema Mihrshahi, and Sara Grafenauer   117

7   Fostering Student Leadership: An International Student
    Challenge to Address SDG2 Zero Hunger
    Karen Oberer, Jolynn Shoemaker, and
    Thomas Rosen-Molina   137

    Winning Student Team Case Study: End-Hunger
    Community Center: A Collaborative Student Idea to
    End Hunger
    Davrina Rianda, Edivan Anjo Ramos, Hafisat Oladimeji,
    Privillege Muleya, Tanaka Murambi, Eviana Barnes   148

8   A Compilation of Global Cases on Teaching,
    Learning, and Campus Stewardship   161

    Student Competitions
    Triggering Change – An SDG2 Challenge Competition
    hosted by John Cabot University in Rome
    Michèle Favorite and Silvia Carnini Pulino   162

    Love Student Leftovers A Digital Student Cooking
    Competition in the UK and Ireland
    Karen Cripps, Pariyarath Sangeetha Thondre
    and Jo Feehily   166

Curriculum Innovations
Zero Hunger Training for English Language Teacher
Candidates in Turkey
Ilknur Bayram and Özlem Canaran   170

Sustainable Food Systems: A Live Event for Accounting
Students at a University in Northern Ireland
Xinwu He   173

Meeting Development Goals in Education:
An Interdisciplinary Approach Focused on
Food at the University of York, United Kingdom
Tim Doheny-Adams, Ulrike Ehgartner, James Stockdale   176

*Campus Projects And Curriculum Addressing
Hunger In The Global North*
Soul Food: How Two Canadian Students' Legacy
Saved 100,000 Pounds of Food
Matt Hopley   180

When Hunger is in Your Higher Education Classroom
in the United States
Xenia K. Morin   182

About the Editors   189

About the Contributors   191

LIST OF FIGURES AND TABLES

FIGURES

Fig. 2.1	Vegetable Plants Grown on the Terrace/Rooftop Garden.	32
Fig. 2.2a	High-technology Supported Urban Agriculture in the Suburbs.	33
Fig. 2.2b	Traditional Urban Agriculture in the Suburbs.	34
Fig. 2.3	Orchid Farming.	35
Fig. 3.1	Treatment and Control Group Productivity Means from 2014 to 2017 of Sugar Companies in Peru.	54
Fig. 5.1	Separate File.	92
Fig. 6.1	University of New Mexico (USA) Campus Community Garden.	123
Fig. 7.1	International Student Challenge Timeline.	140
Fig. 8.1	Instagram Content from the Competition.	167
Fig. 8.2	Students Interviewing Civic Partners During a Leftover 'Cook-off' Challenge on Campus.	168
Fig. 8.3	Phases of the Project.	171
Fig. 8.4	Live Event Plan.	175

TABLES

Table 1.1	SDG2 Outcome Targets.	3
Table 2.1	Ability of Urban Agriculture to Meet Demand in HCMC.	25
Table 2.2	Urban Residents' Perceptions Regarding Urban Gardens.	27

Table 2.3   Biomass Productivity and CO_2 Absorption
                Capacity of Produce Grown in the
                Trial Greenhouse.   28
Table 3.1   Sugar Company Description, Location, and
                Economic Activity.   51
Table 3.2   Productivity – Produced Tonnes (Bag of 60 kg)
                of Sugar per Labour Unit.   52
Table 3.3   Difference-in-Differences Estimation Model
                Results.   55
Table 5.1   Synergy Between MW2063 and Agenda
                2030 for SDGs.   90
Table 5.2   Summary of Research Conducted at LUANAR
                on Improving the Safety, Nutrient Quality,
                and Acceptability of Complementary Foods.   102
Table 7.1   Judging Rubric for the SDG-2 International
                Student Challenge.   145
Table 7.2   Example of LFC Benefits.   150

SERIES EDITOR PREFACE
Professor Wendy Purcell PhD FRSA

Higher education (HE) makes an important contribution to realising the sustainable development goals (SDGs). Teaching and learning support the development of responsible citizens as scholars, leaders, entrepreneurs, and professionals. Curiosity-driven and socially impactful research and innovation help advance knowledge frontiers and find solutions for the world's most pressing issues. As anchor institutions, universities and colleges are also active in civic and community settings, working in partnership with other stakeholders. Given the fierce urgency of (un)sustainable development, the climate crisis, and widening inequity within countries and across the globe, HE institutions (HEIs) need to do more and go faster to deliver fully on their potential to help achieve the SDGs.

This book series focuses on the role of HE in advancing the SDGs, identifying some actionable and scalable initiatives and pointing to opportunities ahead. In sharing the ways and means universities and colleges across the world are engaging with the SDGs, the series seeks to both inspire and enable those in the HE sector and stakeholders beyond to transform what they do and how they do it and thereby hasten progress towards Agenda 2030. Insights gleaned from case studies, reflective accounts, and student stories can help the HE sector both deepen and accelerate its engagement with the SDGs. Each book seeks to capture examples of how HEIs are fulfilling delivery of their academic mission *and* progressing the SDG concerned. Illustrating the work of students, that undertaken by faculty and staff of the institution, and conducted with partners, positions HE as a change agent operating at a systems level to help create a world that leaves no one behind.

This book on HE and SDG2 highlights the work of universities and colleges focused on achieving the goal of Zero Hunger – to end hunger, achieve food security, improve nutrition, and promote sustainable agriculture. Hunger is a complex problem that reflects the various scales from global to hyperlocal, from the community to the individual, and from field to fork, and it is growing at an alarming rate. This book clearly articulates differing perspectives across the world and serves to challenge some of our assumptions, for example that food insecurity and promoting sustainable agriculture are not simply concerns in the global South but are rapidly emerging as major issues for affluent nations of the global North. It does this by adopting a truly international flavour with authors drawn from Australia, Canada, China, Indonesia, Italy, Malawi, Northern Ireland, Peru, Turkey, the United States of America, the United Kingdom, Vietnam, and Zimbabwe. The book reveals how local issues relating to SDG2 are translated into the activities and mission of HEIs and serve to foster global partnerships.

SDG2 reflects undernourishment, chronic and acute malnutrition, wasting and stunting, as well as food insecurity and hidden hunger. It interfaces with so many of the other SDGs, 12 of the 17 goals from good health and well-being (SDG4) and quality education (SDG4) to climate change (SDG13) and life below water (SDG14) and on land (SDG15) as well as those relating to equity (SDG5 gender equality and SDG10 reduced inequalities). Food security is achieved only when everyone has physical, social, and economic access to sufficient, safe, and nutritious food that meets their dietary needs and food preferences for an active and healthy life. From the use of high-technology supported agriculture models focused on urban food security (Chapter 2), to international research on edible insects as a novel protein source (Chapter 4), the book highlights the work of HEIs in stewardship of land as well as on-campus gardens (Chapter 6), with case studies that can be copied and scaled by the HE sector.

Health of people, planet, and shared prosperity rely upon the full participation of HE with universities and colleges, in turn, needing to pursue greater engagement with the SDGs – not least to reduce their own environmental footprint and become more equitable.

As organisations that have stood for many centuries in some cases, this demands that they adapt with new models of learning, research partnerships, and leadership and governance frameworks. Immersive engagement with the SDGs can catalyse pedagogic innovation, serve to refresh curricula, and stimulate new programme development. It can also open new avenues for research, attract new sources of funding, and energise people to deliver on the academic mission. HEIs can play a critical role in developing new systemic and transformative solutions through interdisciplinary and multistakeholder collaboration and a purposeful focus on the SDGs. This book illustrates this approach as it relates to SDG2 with HEIs bringing their key assets of curiosity and the pursuit of knowledge and its application to partners seeking solutions and driving innovation, operating in both local and global networks, and connecting the worlds of learning, work, and entrepreneurship in support of more sustainable development. Sustainability is a goal for today and sustainable development an organising principle for universities and colleges.

ACKNOWLEDGEMENTS

We extend our deep gratitude to Professor Wendy Purcell for her invaluable guidance and mentorship throughout. From supporting a vision of the book through to initiating contacts and shaping book structure, Wendy has been a constant source of wisdom and inspiration. We are also indebted to Katy Mathers and the team at Emerald along with reviewers, for their enthusiastic support and advice.

It goes without saying that we would not be in the privileged position to be writing this without the authors and student contributors. It has been a pleasure to see individual chapters weave together into this collaborative shared story of the excellent work that is happening globally to address SDG2 Zero Hunger. We are truly humbled by everyone's work and have valued the opportunity to work across disciplines on a global scale.

Personally, Karen dedicates this book to her sons Dylan and Brandon and hopes that the world they grow up in sees much faster progress towards eliminating food insecurity, hunger, and malnutrition. Sangeetha thanks her family, friends, and colleagues who have inspired and supported throughout her career, and Karen for her leadership and enthusiasm which made this project a very fulfilling experience.

1

AN INTRODUCTION TO SDG2 ZERO HUNGER

Karen Cripps and Pariyarath Sangeetha Thondre

Oxford Brookes University, UK

INTRODUCTION

The United Nations *Sustainable Development Goal2* (SDG2) is to 'End hunger, achieve food security and improved nutrition and promote sustainable agriculture'. The global challenge of hunger and food insecurity is exacerbated by a combination of factors including the Covid-19 pandemic, conflict, climate change, and deepening inequalities. This goal has various dimensions, namely, social in aiming to end hunger, environmental/climate through sustainable agriculture, and economic concerning food security and nutrition. Despite the launch of the *Sustainable Development Goal* in 2015, the number of people facing hunger and food insecurity continues to rise with a recent United Nations (2023) report describing it as a 'polycrisis' given alarming statistics that:

> In 2022, about 9.2 per cent of the world population was facing chronic hunger, equivalent to about 735 million people – 122 million more than in 2019. An estimated

29.6 per cent of the global population – 2.4 billion people – were moderately or severely food insecure, meaning they did not have access to adequate food. This figure reflects an alarming 391 million more people than in 2019 (United Nations, 2023, p. 14).

At the most recent United Nations Climate Change Conference (COP28), world leaders highlighted the criticality of building a stronger food system for generations to come that is resilient to climate and humanitarian shocks. The challenges faced by countries in the Global North and Global South are very different with respect to SDG2 since the former may be concerned with nutritional security and sustainability of food systems generally, while the latter may be facing widespread food insufficiency and malnutrition, alongside the other concerns. However, in Western nations such as the United Kingdom, 'average' situations are said to hide huge variations according to socio-economic status and struggling food systems that paradoxically need to address diet-related diseases alongside food insecurity among vulnerable populations (United Nations Global Compact Network, 2023).

SDG2 is composed of 8 targets (5 outcome and 3 implementation-related) and 14 indicators. This book is appropriately timed at the mid-point of the 2030 agenda to consider the role of higher education in addressing the five outcome-related targets on ending hunger and ensuring the establishment of a sustainable food system in a world experiencing climate challenges (see Table 1.1). Given debates surrounding the operationalisation of targets and indicators pertinent to SDG2 are comprehensively addressed by authors such as Cheo and Tapiwa (2021), this book seeks to explore and showcase the contributions of *higher education institutions* (HEIs) to this global goal. The various chapters validate UNESCO-ISEALC's (2023) bold and ambitious statement that 'Higher education institutions can contribute to the eradication of hunger'. This book includes leading examples of HEIs as 'change agents', showing that many do, indeed, 'walk the talk' (Kumar et al., 2024). If the premise is adopted that 'none of the SDGs will be fully achieved without the contribution of the university sector' (SDSN, 2020), then a world with 'zero hunger' is achieved when

Table 1.1.   SDG2 Outcome Targets.

SDG2 Target	Key Principles	Indicative Measurements
2.1	Hunger Food security Safe, nutritious food	Food insecurity experiences Undernourishment
2.2	Malnutrition	Stunting, wasting, anaemia
2.3	Agricultural productivity and income	Volume of production by enterprise size Income of small-scale producers by gender and indigenous status
2.4	Sustainable food (production) systems; climate/disaster resilience	Agricultural areas under productive and sustainable agriculture
2.5	Genetic diversity Traditional knowledge	Plant and animal genetic resource conservation

Source: Authors based on United Nations (2015).

the transdisciplinary projects presented here are adopted widely and scaled globally.

A STRUCTURE TO RECOGNISE APPROACHES TAKEN BY THE GLOBAL SOUTH AND GLOBAL NORTH

The global, transdisciplinary nature of SDG2 demands recognition of the differing priorities, environments, and socio-economic contexts in which HEIs are located. Kumar et al. (2024, p. 26) contend that

> *While the Global North is focused on 'de-growth' in the context of a significantly improved standard of living, much of the Global South is struggling to develop through economic growth strategies, thus focusing particularly on SDGs 1–3 (eradication of poverty, zero hunger and good health).*

Accordingly, as this book took shape, a clear pattern emerged by which contributors from the Global South could be grouped on the

basis of Times Higher Education (THE) (2023) SDG Impact Rankings as 'research'-led policy and 'community outreach' initiatives
(Part 1), and contributors from the Global North could be linked
to a 'teaching and learning', 'stewardship' (campus projects), and
'community outreach' initiatives (Part 2).

Ashida's (2023) analysis of the role of HEIs in achieving the
SDGs recommends global collaboration with a sense of empathy
to understand the lived experiences of those people affected by the
foci of SDG2 targets. It can be contended that this book provides
illustrative examples of such collaboration, with projects that most
certainly nurture empathy among the actors involved that will
hopefully trigger further empathetic insight and knowledge building through sharing them here.

Featuring 37 contributing authors from Indonesia, Vietnam,
Malawi, Zimbabwe, Peru, China, Australia, Northern Ireland,
Canada, the United States, Turkey, Italy, and the United Kingdom, the book represents geographically diverse examples of HEI
approaches to SDG2. Within the chapters and cases, the student
voice is featured throughout, from undergraduate through to PhD
research students, alongside faculty members with research and/or
pedagogic expertise.

Part 1 comprises five chapters that cover research-based policy
and community perspectives offering insights into policy perspectives and the criticality of concerns in low- and middle-income
countries facing arguably the most widespread severe impacts
of hunger, food insecurity, malnutrition, and unsustainable food
systems. Vu Thi and Juntti (Chapter 2) present innovative, high
technology solutions to support urban agriculture for food security
in Ho Chi Minh City based on findings from an interdisciplinary
collaboration between HEIs in the United Kingdom and Vietnam.
Using an example of the sugar industry in Peru, Padilla and
Henríquez (Chapter 3) demonstrate the effect of adverse weather
conditions on agricultural productivity and the resultant economic
impact. Chopera et al. (Chapter 4) highlight the importance of
using edible insects to provide culturally acceptable and nutritious food for malnourished children in Zimbabwe and is another
example of North–South collaboration. Chapter 5 illustrates the
contributions of a university in Malawi towards the achievement

of SDG2 targets by promoting the use of underutilised crops and improving their nutritional value by biofortification. A case study from China is also included at the end of Chapter 5 to depict strategies adopted by urban Chinese residents to ensure food security.

Part 2 is composed of three chapters that cover teaching and learning and governance through campus outreach initiatives and provides insights into the role that higher education plays in raising awareness and skills building. This begins with Lin et al.'s analysis (Chapter 6) of the potential for campus food gardens to achieve food literacy and security in Australia. Oberer et al. (Chapter 7) describe a case of a global SDG2 challenge led by academic teams in the United States and Canada with the winning entry from an internationally dispersed group of students working on a solution to nutrition and food insecurity in Indonesia.

The final chapter of the book (Chapter 8) is a series of seven case studies that directly contribute to the core themes of SDG2 targets and reflect the breadth of curriculum-based initiatives in the Global North that address both national and global issues. These are grouped into three sections beginning with student competitions. Favorite's case study of an SDG2 challenge competition involves cross-disciplinary European students working in teams with input from sustainable food system experts, which like Cripps et al.'s case study of a competition based in the United Kingdom in partnership with an Indian university highlights the close inter-connections between SDG2 and SDG12 in consideration of food waste.

The second section of this chapter presents curriculum innovation case studies by Bayram and Canaran's exposition of embedding SDG2 within a teacher training course in Turkey, and He's illustration of how it is embedded in an accountancy module at a university in Northern Ireland. Doheny-Adams et al. illustrate a pioneering interdisciplinary module open to students across the University of York that examines the 'future of food' from diverse perspectives. The final section of this chapter includes cases that illustrate campus projects and curriculum approaches focusing on addressing hunger, beginning with Hopley's account of a Canadian student initiative to distribute food to local vulnerable populations and Morin's account of how the American curriculum responds to national level strategy on addressing food insecurity. This is a fitting chapter to close the book, as it leaves the reader with a sense of how everyone can play a part in awareness and action for SDG2.

HOW THE CHAPTERS ADDRESS SDG2 TARGETS

Food Security (2.1)

With more than 600 million people predicted to be facing hunger worldwide by 2030, achieving SDG2 targets 2.1 and 2.2 requires a multisector, coordinated approach to transform the food system, health system, and social protection system (FAO, IFAD, UNICEF, WFP and WHO, 2023). Innovative and sustainable approaches are needed to utilise traditional practices and exploit technological advances to improve food security in Africa and Asia, where the highest proportion of individuals are affected.

Several contributions are made that highlight the challenge of food insecurity faced in the Global South. Chopera et al. (Chapter 4) and Mwangwela et al. (Chapter 5) present the current state of food insecurity in Sub-Saharan Africa caused by climate change and environmental and socioeconomic factors. Vu Thi and Juntti (Chapter 2) allude to urban food insecurity in Vietnam contributed by urban migration and loss of agricultural land. Jiahui and Guojun's case study (within Chapter 5) examines dietary practices of urban residents in a region of China as part of a broader global study that examines dietary practices of city residents in building reflections and joint solutions. The Chinese case is especially concerned with issues around food safety. Through qualitative interviews, this reveals the relationship between dietary choices and factors such as attachment to local/regional foods, trustworthy, and accessible retail opportunities.

Interestingly, many contributions also address food insecurity in the Global North. Lin et al. (Chapter 6) showcase how universities can adopt stewardship strategies through campus food gardens that contribute to building food literacy and enabling food security. This is especially meaningful in demonstrating the challenge of food security and hunger among Western populations, where such issues might not necessarily be considered. Chapter 6 highlights, through survey research into student populations in Australia and the use of campus food gardens across curriculum disciplines, the links between food security and student wellbeing and performance outcomes. Morin's case study (Chapter 8) examines similar challenges in the United States and the role of educators in stimulating discussion with students around food security, while Hopley's case study (Chapter 8) also highlights challenges of food insecurity in the region of Ontario, Canada. This is a fascinating exemplar of university stewardship through a volunteer student-led initiative to deliver surplus canteen food to vulnerable residents living in shelters.

(Mal) Nutrition (2.2)

Despite global efforts to address malnutrition and food insecurity, the share of the global population experiencing undernourishment accounts for 9.3% (Our World in Data Team, 2023). This figure has been increasing since 2015, predominantly due to the prevailing impact of conflict, exacerbating effects of climate change, and the lingering shock of the Covid-19 pandemic. A resulting outcome of this polycrisis situation is the global prevalence of chronic and acute forms of malnutrition affecting children under five. The 2023 figures show 45 million and 148 million children under 5 are affected by wasting and stunting respectively (United Nations Children's Fund (UNICEF), World Health Organization (WHO), International Bank for Reconstruction and Development/The World Bank, 2023).

The chapters by Chopera et al. (Chapter 4) and Mwangwela et al. (Chapter 5) reflect these statistics on malnutrition, supporting the high prevalence of stunting, wasting, and micronutrient deficiencies in children and adults of sub-Saharan Africa. They identify various projects where the addition of plant- and insect-based foods addresses multiple goals fostering gender inclusion, traditional cuisine, cultural value, and sustainability. Oberer et al. (Chapter 7) discuss a global SDG Zero Hunger Consortium that runs an annual challenge competition. The most recent winning student team from Indonesia show how their proposed 'End-Hunger Community Centre' would address challenges of access to health and nutrition by providing diet and nutrition knowledge and enabling local empowerment through home growing and support to local food producers. The students' case study demonstrates the impact a global team of students brought together through the forum of a competition can have in driving both awareness and capacities to develop solutions to global challenges.

Agriculture Productivity (2.3)

At a time when some parts of the world are undergoing rapid urbanisation and economic development, the crippling effect of war, conflict, and civil war affecting many countries (e.g. Afghanistan, Colombia, Ethiopia, Iraq, Libya, Mali, Myanmar, Russia, Somalia, Syria, the Central African Republic, the Democratic Republic of Congo, Ukraine, and Yemen) continue to increase the risk of food insecurity (ACLED, 2024). In an already fragile world affected by Covid-19, war has further disrupted agriculture productivity, food exports/imports, supply chains, and food, oil, fuel, and fertiliser markets, hindering progress towards achieving SDG2 targets 2.3 and 2.4 (Ben Hassen & El Bilali, 2022). Critically, many African countries depend on Russia and Ukraine for fuel and food imports, and the ongoing war has huge implications for progress towards SDG2 in Africa (Mhlanga & Ndhlovu, 2023).

The impact of war on increasing food prices is most felt in regions of sub-Saharan Africa, where perpetuating climate catastrophes, locust swarms, civil wars, and crop failures contribute to food insecurity (Mhlanga & Ndhlovu, 2023). The impacts of climate change on agricultural productivity are elucidated in Chapter 3 as Padilla and Henriquez highlight the negative consequences of the El Niño phenomenon in the South American region, elucidating the challenges posed by weather patterns in achieving double agriculture productivity by 2030. Particular attention is drawn to the disruption caused by heavy rain to the sugar industry, schools, and HEIs, slowing down economic growth in Peru. The authors call for collaboration between HEIs, governments, and industries to build resilience against climate change impact on productivity.

Sustainable and Climate-Resilient Food (2.4)

Reduced food exports from war-affected countries may lead to intensive agriculture elsewhere to meet the global demand for food, thus negatively affecting the achievement of SDG2.4 targets (Pereira et al., 2022). As the global agrifood systems attempt to feed the expanding global population in the coming years, one of the major challenges is to do it without compromising the 1.5°C global warming target (FAO, 2023b). To achieve this, FAO presented a global road map at COP28 aligning climate and agrifood systems solutions, calling for breaking down silos and adopting a more holistic strategy through international collaboration and coordination (FAO, 2023a). Simultaneously, the Stockholm Resilience Centre reports that six out of nine Earth's planetary boundaries have exceeded safe operating limits, highlighting the need for an agrifood system transformation to mitigate its effects on planetary health (Richardson et al., 2023).

Climate resilience is becoming ever more important in this century as humanity grapples with extreme temperatures, drought, floods, and wildfires: many in places where they are least expected and affecting the ability of farmers in rural areas to grow crops. Concurrently, more than 50% of the global population currently lives in urban areas (FAO, IFAD, UNICEF, WFP and WHO, 2023) prompting a rethink about the way food is grown. Considering the cost-of-living crisis following the Covid-19 pandemic and an increasingly mobile global population due to conflicts and associated challenges, even countries in the Global North are experiencing increasing levels of food poverty in this decade, forcing many to rely on unhealthy foods to satisfy hunger (Goudie, 2023). These practices are predicted to drive an increase in overweight/obesity and associated chronic diseases. In this context, it is pertinent to think beyond rural farms and redesign our cities and streets to grow fresh food while encouraging innovative methods such as vertical farming, hydroponics, aeroponics, and aquaponics, which are now being widely considered in the Global North (Yuan et al., 2022).

Urban gardens are vital in changing food consumption patterns towards diets with low carbon footprint and promoting pro-environmental behaviours including reductions in waste generation, food transport, and use of packaging (Puigdueta et al., 2021). Vu Thi and Juntti's chapter (Chapter 2) provides sustainable and technology-driven solutions to achieve food security in rapidly growing Ho Chi Minh City. They demonstrate the possibility to create a resilient and modern urban agriculture model to promote food production, supported by the willingness of citizens to use high-tech agriculture systems. Ho Chi Minh City's urban residents set a great example for the rest of the world by embracing modern agriculture practices to create a circular and low carbon economy rather than relying on processed foods.

Genetic Diversity (2.5)

While it is important to discuss the visible effects of food insecurity manifesting in the forms of wasting, stunting, and overweight/obesity; 'hidden hunger' in the form of micronutrient deficiencies is equally relevant in the context of Global North and Global South (Bailey et al., 2015; Stevens et al., 2022). Utilising plant genetic resources by biofortification methods to address these deficiencies and their effects on malnutrition is promising and offers hope for making progress with SDG2 targets (Bouis, 2018). The use of these methods in combination with nutrition education to promote more diverse diets offers sustainable solutions to address hidden hunger at the global level (Van Der Straeten et al., 2020). Modern technologies such as genome editing also have the potential to contribute to the SDG2 targets (Smyth, 2022) and HEIs worldwide play a fundamental role in driving progress towards the 2030 agenda through knowledge creation, innovation, skills generation, and fostering collaboration.

Two chapters in this book portray the efforts of HEIs in Africa in accelerating the achievement of SDG2 by 2030. In Chapter 4, Chopera et al. describe the challenges faced by the African continent in addressing food and nutrition security. Through their case study on creating a value-added product using edible insects, they look beyond plant and animal genetic resources for food (referred to in SDG2.5) to end all forms of malnutrition and deficiencies in Zimbabwean children. Mwangwela et al. (Chapter 5) exemplify the research conducted in an Agriculture and Natural resources university in Malawi to ensure the quality and safety of several plant-based food sources. Their gender inclusive research and efforts to promote underutilised legumes and biofortified crops are instrumental in ending malnutrition in Africa while enhancing health equality.

RECOMMENDATIONS

Through the current leading global framework of university action for the SDGs, the THE (2023) Impact Rankings measure teaching and learning, research, outreach, and stewardship, this book showcases exemplars of HEIs as 'change agents' that 'walk the talk' (Kumar et al., 2024). Highlighting examples of good practice is especially important considering research into 363 universities by the International Association of Universities revealed little institutional work is being done on SDG2 (Toman et al., 2023). Awareness and understanding underpin action in progressing the SDGs, and our analysis through the collation of good practices for this book, indicates some areas of potential confusion in measurements and ranking tools.

We do not claim an exhaustive analysis of global mechanisms that underpin HEI approaches to SDG2, but rather focus analysis here on two leading frameworks. The THE impact rankings methodology for assessment of SDG2 includes metrics on research (through publications) related to hunger (defined as 'a severe lack of food, causing suffering or death'), campus food waste, student hunger (student food insecurity, food choices, and interventions to target hunger), proportion of graduates who receive a degree associated with 'any aspect of food sustainability within an agricultural or aquacultural course', and efforts to address hunger at a national level (encompassing initiatives based on knowledge/skills/technology to local farmers and food producers). It would be helpful if the ratings were mapped to the SDG targets and indicators to facilitate greater depth of insight into university approaches.

To enable a more systematic consideration of progress, greater synthesis is required in HEI SDG assessment/reporting frameworks that are linked to SDG2 targets. A clearer mapping against SDG2 targets in HEI tools would help maximise the utility and comparability of diagnostic assessment tools such as the open resource Sustainability Evaluation Tool for Higher Education Institutions

(SET4HEI). 'Education for the Sustainable Development Goals' (SDSN, 2020) guidance classifies research, education, operations and governance, and external leadership that can be aligned to THE and SET4HEI. Yet, it is the detail on the targets that hinders comparability. This is concerning, since as graduates transition into employment, a clear focus on the targets underpins effective organisational action and measurement also. A review of university reports and website communications on SDG2 reveals diverse and immensely impressive initiatives, but we consider they would be strengthened if better aligned to the targets. If the premise is adopted that 'none of the SDGs will be fully achieved without the contribution of the university sector' (SDSN, 2020) then a world with 'zero hunger' demands target-level communication and reporting.

SDG2 is complex in many ways; it is not just about access to adequate food for the global population. Ensuring that the food available is nutritious, affordable, and sustainable is a challenge that high- and low-income countries alike are struggling to overcome in this decade of economic slowdown. At the time of establishing SDG2, the world had not experienced the level of disruption of recent years. Conflict is by far one of the main reasons for the increase in hunger and malnutrition since 2015 (FAO, IFAD, UNICEF, WFP and WHO, 2023). It not only destroys natural resources used to grow and produce food but also targets places where food is stored, sold, transported, and imported. The challenges in the journey to end global hunger demands interdisciplinary and collaborative efforts from the current and future workforce, governments, policy makers, businesses, and academics.

As captured in this book, HEIs are using innovative teaching, learning, research, and co-curricular activities across multiple disciplines to create awareness and action to tackle food insecurity. It is possible for HEIs to evaluate the impact of war on SDG2 targets using modelling and predictions (Ben Hassen & El Bilali, 2022; Mhlanga & Ndhlovu, 2023); however, short-term humanitarian relief must be integrated with long-term development and disaster risk reduction to attain SDG2 in conflict-affected areas. While this book includes chapters reflecting all of the SDG2 targets, it is acknowledged that there is less emphasis on SDG2.5 on maintaining genetic diversity of plants and animals. Hunger is a complex

problem interlinked with poverty, education, equity, health and
well-being, employment, water and sanitation, and climate change
such that achieving SDG2 depends on the progress with these other
SDGs. The resounding message of this book is that continued systemic, interdisciplinary, trans-global, target-driven SDG initiatives
across teaching and learning, research, stewardship, and community outreach are central to the pursuit of zero hunger.

REFERENCES

ACLED. (2024). *The armed conflict location & event data project.* https://
acleddata.com/conflict-index/

Ashida, A. (2023). The role of higher education in achieving the sustainable
development goals. In S. Urata, K. Kuroda, & Y. Tonegawa (Eds.),
Sustainable development disciplines for humanity (pp. 71–84). Springer.

Bailey, R. L., West, K. P., Jr., & Black, R. E. (2015). The epidemiology of
global micronutrient deficiencies. *Annals of Nutrition and Metabolism*,
66(Suppl 2), 22–33. https://doi.org/10.1159/000371618

Ben Hassen, T., & El Bilali, H. (2022). Impacts of the Russia-
Ukraine war on global food security: Towards more sustainable
and resilient food systems? *Foods*, *11*(15), 2301. https://www.mdpi.
com/2304-8158/11/15/2301

Bouis, H. (2018). Reducing mineral and vitamin deficiencies through
biofortification: Progress under HarvestPlus. *World Review of Nutrition
and Dietetics*, *118*, 112–122. https://doi.org/10.1159/000484342

Cheo, A. E., & Tapiwa, K. A. (2021). *SDG2 – Zero hunger: Food security,
improved nutrition and sustainable agriculture.* Emerald Publishing.

FAO (2023a). Achieving SDG 2 without breaching the 1.5°C threshold:
A global roadmap, Part 1 – How agrifood systems transformation
through accelerated climate actions will help achieving food security and
nutrition, today and tomorrow, In brief. Rome, Italy. https://www.fao.org/
interactive/sdg2-roadmap/en/

FAO (2023b). *COP28 spotlights agrifood systems' potential to address
climate impacts and achieve 1.5 °C goal.* https://www.fao.org/newsroom/

detail/cop28-fao-spotlights-agrifood-systems-potential-to-address-climate-
impacts-and-achieve-1.5-c-goal/en

FAO, IFAD, UNICEF, WFP and WHO. (2023). *The state of food security
and nutrition in the World 2023. Urbanization, agrifood systems
transformation and healthy diets across the rural–urban continuum.*
Rome, FAO. https://doi.org/10.4060/cc3017en

Goudie, S. (2023). *The Broken Plate 2023: The state of the nation's food
system.* https://foodfoundation.org.uk/publication/broken-plate-2023

Kumar, P., Caporarello, L., & Agrawal, A. (2024). Research from
the Global North and South on HEIs as a driver for the SDGs. In
L. Caporarello, P. Kumar, & A. Agrawal (Eds.), *Higher education for
the sustainable development: Bridging the Global North and South*
(pp. 26–35). Emerald Publishing.

Mhlanga, D., & Ndhlovu, E. (2023). The implications of the Russia–
Ukraine war on sustainable development goals in Africa. *Fudan Journal
of the Humanities and Social Sciences*, 16(4), 435–454. https://doi.
org/10.1007/s40647-023-00383-z

Our World in Data Team. (2023). *End hunger, achieve food security
and improved nutrition and promote sustainable agriculture.* https://
ourworldindata.org/sdgs/zero-hunger

Pereira, P., Zhao, W., Symochko, L., Inacio, M., Bogunovic, I., & Barcelo, D.
(2022). The Russian-Ukrainian armed conflict will push back the
sustainable development goals. *Geography and Sustainability*, 3(3),
277–287. https://doi.org/10.1016/j.geosus.2022.09.003

Puigdueta, I., Aguilera, E., Cruz, J. L., Iglesias, A., & Sanz-Cobena, A.
(2021). Urban agriculture may change food consumption towards low
carbon diets. *Global Food Security*, 28, 100507. https://doi.org/10.1016/
j.gfs.2021.100507

Richardson, K., Steffen, W., Lucht, W., Bendtsen, J., Cornell, S. E., Donges,
J. F., Drüke, M., Fetzer, I., Bala, G., Von Bloh, W., Feulner, G., Fiedler,
S., Gerten, D., Gleeson, T., Hofmann, M., Huiskamp, W., Kummu, M.,
Mohan, C., Nogués-Bravo, D., … Rockström, J. (2023). Earth beyond six
of nine planetary boundaries. *Science Advances*, 9(37), eadh2458.

SDSN. (2020). *Sustainable development solutions network – Accelerating education for the SDGs in universities*. https://www.unsdsn.org/accelerating-education-for-the-sdgs-in-univerisities

Smyth, S. J. (2022). Contributions of genome editing technologies towards improved nutrition, environmental sustainability and poverty reduction. *Frontiers in Genome Editing*, 4, 863193. https://doi.org/10.3389/fgeed.2022.863193

Stevens, G. A., Beal, T., Mbuya, M. N. N., Luo, H., Neufeld, L. M., Addo, O. Y., Adu-Afarwuah, S., Alayón, S., Bhutta, Z., Brown, K. H., Jefferds, M. E., Engle-Stone, R., Fawzi, W., Hess, S. Y., Johnston, R., Katz, J., Krasevec, J., McDonald, C. M., Mei, Z., … Young, M. F. (2022). Micronutrient deficiencies among preschool-aged children and women of reproductive age worldwide: A pooled analysis of individual-level data from population-representative surveys. *The Lancet Global Health*, 10(11), e1590–e1599. https://doi.org/10.1016/S2214-109X(22)00367-9

Times Higher Education. (2023). *Impact rankings 2023: Zero hunger (SDG2) methodology*. https://www.timeshighereducation.com/impact-rankings-2023-zero-hunger-sdg-2-methodology

Toman, I., van't Land, H., & Harris, M. (2023). *International association of universities – Accelerating action for the SDGs in higher education*. https://iau-aiu.net/IMG/pdf/iauhesdsurvey2023_accelerating_actions.pdf

UNESCO-ISEALC. (2023). *SDG2: Zero hunger*. https://campus.iesalc.unesco.org/encuesta/execert/20/

United Nations (2023). *The sustainable development goals report*. https://unstats.un.org/sdgs/report/2023/

United Nations Children's Fund (UNICEF), World Health Organization (WHO), International Bank for Reconstruction and Development/The World Bank. (2023). *Levels and trends in child malnutrition: UNICEF/WHO/World Bank Group Joint Child Malnutrition Estimates: Key findings of the 2023 edition*. UNICEF and WHO.

United Nations Global Compact Network. (2023). *Measuring up 2.0: How the UK is measuring up on the sustainable development goals*. https://www.unglobalcompact.org.uk/measuring-up/

United Nations. (2015). *Global indicator framework for the sustainable development goals and targets of the 2030 agenda for sustainable development.* https://unstats.un.org/sdgs/indicators/Global%20Indicat or%20Framework%20after%202022%20refinement_Eng.pdf

Van Der Straeten, D., Bhullar, N. K., De Steur, H., Gruissem, W., MacKenzie, D., Pfeiffer, W., Qaim, M., Slamet-Loedin, I., Strobbe, S., Tohme, J., Trijatmiko, K. R., Vanderschuren, H., Van Montagu, M., Zhang, C., & Bouis, H. (2020). Multiplying the efficiency and impact of biofortification through metabolic engineering. *Nature Communications*, *11*(1), 5203. https://doi.org/10.1038/s41467-020-19020-4

Yuan, G. N., Marquez, G. P. B., Deng, H., Iu, A., Fabella, M., Salonga, R. B., Ashardiono, F., & Cartagena, J. A. (2022). A review on urban agriculture: Technology, socio-economy, and policy. *Heliyon*, *8*(11), e11583. https://doi.org/10.1016/j.heliyon.2022.e11583

Part 1

THE GLOBAL SOUTH RESEARCH-BASED POLICY AND COMMUNITY PERSPECTIVES

2

URBAN RESILIENCE FROM AGRICULTURE: A CASE STUDY OF HO CHI MINH CITY IN VIETNAM

Quyen Vu Thi[a] and Meri Juntti[b]

[a]Van Lang University, Vietnam
[b]Middlesex University, UK

ABSTRACT

This chapter focuses on the potential of urban agriculture to support progress in SDG targets 2.1, 2.2, 2.3, and 2.4 in Ho Chi Minh City (HCMC), Vietnam. The chapter integrates findings from the British Council-funded project, 'Urban Resilience from Agriculture through Highly Automated Vertical Farming in the UK and Vietnam', undertaken in collaboration with Middlesex University, Van Lang University, and local agricultural stakeholders in HCMC. Food security in the city faces multiple challenges ranging from significant in-migration, decreasing area of cultivated land, the impact of the Covid-19 pandemic that continues to depress the economy and disrupt food supply chains, and climate change impacts affecting the environment and people throughout the city. HCMC accommodates a substantial agricultural sector, which is evolving from traditional to modern production practices. City's leaders established numerous policies that emphasise green, circular economies, climate change resilience, and low carbon emissions fuelling demand for agricultural solutions that integrate traditional and modern technologies that can be embedded in the local topography, soil types, architectural space, and native culture. Findings from greenhouse trials, community awareness surveys, and stakeholder-led workshops point to a range of high-technology-supported agriculture models that, if applied flexibly throughout the varying context of the urban area, have good scope to help Ho Chi Minh City and meet its growing need for food as well as its sustainability aspirations.

Keywords: Vietnam; Ho Chi Minh city; urban agriculture; vertical farming; high-technology production systems; urbanisation; urban food security

1. INTRODUCTION

Vietnam is a fast-developing lower middle-income country. In 2019, it ranked 117th out of 189 countries for the human development index (UNDP, 2022). Three decades of economic and social progress render it well poised to address the United Nations sustainable development goals (SDGs). Among its Asian peers, Vietnam's progress on the SDGs is second only to Thailand. Globally, Vietnam was 49th out of 166 countries in 2020, rising by four places from 2019 (Ministry of Planning & Investment, 2021). While Goal 2: zero hunger is one of the five SDGs that Vietnam is on track to complete by 2030, the following targets are posing a challenge: access to safe nutritious and adequate food for all, especially the poor and vulnerable, including the elderly and infants (Target 2.1); reducing all forms of malnutrition, in specific, meeting the nutritional needs of children, adolescent girls, pregnant and lactating women, and the elderly (Target 2.2); increasing labour productivity in agriculture and income of agricultural workers (Target 2.3); and ensuring sustainable food production and applying resilient agricultural production methods that increase productivity and output, maintain ecosystems, and enhance resilience to climate change and other disasters and gradually improve land quality (Target 2.4) (Ministry of Planning & Investment, 2021, p. 12; Ministry of

Vietnam Politics, 2020). These challenges implicate both urban and rural territories, and many are felt particularly keenly in the context of rapidly growing cities, such as HCMC. This chapter discusses the scope for high-technology supported urban agriculture to aid progress in meeting the above SDG targets in the context of HCMC. It draws on the findings from the Urban Resilience from Agriculture project undertaken in collaboration with Middlesex University in the UK, and Van Lang University and agricultural stakeholders in HCMC, Vietnam.

HCMC is one of the five largest cities in Vietnam. It is situated in a tropical climatic region, on the Saigon River above the Mekong River Delta, about 50 miles from the coast of the South China Sea. The city's territory spans highly urbanised areas as well as forests and beaches and integrates a diverse river system. It has a total land area of 209,523 ha (or 2,061 km^2) with 21 districts and 1 urban centre. According to the preliminary census results in 2022, the city's permanent population is 9,166,800 people and, with the temporary population included, rises to about 14 million (HCMC Statistics Department, 2023). According to a report by the Department of Agriculture and Rural Development of HCMC (2022), the entire city has 113,634 ha of agricultural land, accounting for 54.23% of the total land area. The main crops encompass vegetables, flowers, and ornamental plants (accounting for about 5.6% – 6,317 ha).

HCMC is a prospering city. The average income per capita increased from 58.1 million VND/year in 2014 to 53.6 million VND/year in 2022, making the city one of the wealthiest in Vietnam (HCMC Statistics Department, 2023). According to census data, in 2022, there were 39,381 poor and near-poor households, accounting for 1.55% of the total households (HCMC Statistics Department, 2023). The city constitutes a centre of economic activity, culture, training and education, and science and technology and is the largest focal point for international exchange and integration in the country. Benefitting from burgeoning and diverse business and education sectors (6 academies and 50 universities), the city has great potential in technology, management, and international cooperation. The city's leaders are determined to exploit this potential to achieve sustainable development and have affirmed a commitment to developing a green economy, decreasing waste, and reducing greenhouse gas emissions by encouraging low-carbon solutions and the development of circular economic models for integrated and efficient use of outputs from production processes.

Despite economic growth and technological potential, HCMC is not immune to the challenges to food security. A rapidly growing population and the constant influx of rural migrants together with climate change impacts place multiple pressures on agricultural land area and productivity. The city is committed to supporting urban agriculture to enhance food security and food safety, integrate lessons from the Covid-19 pandemic into supply chain management, and respond to climate change, while contributing to a circular and low carbon economy (HCMC People's Committee, 2021a). According to the leader of the Department of Agriculture and Rural Development in HCMC, by 2030, at least 70% of the city's agricultural production area will apply high-technology solutions to increase the value, and efficiency of land use, while creating the premise for building a modern, sustainable urban agriculture sector (Xuan, 2021).

2. THE TRANSITION FROM TRADITIONAL TO MODERN AGRICULTURE AND RESPONDING TO FOOD SECURITY NEEDS IN VIETNAM AND HCMC

2.1. Urban Agriculture in Vietnam

Consequences from the Covid-19 pandemic along with the negative impact of climate change have highlighted the benefits of maintaining a dynamic and creative agriculture sector in the urban context (Langemeyer et al., 2021). In Vietnam, there is a range of popular urban agriculture models to build on. Vu (2022) summarises the most common types as follows:

(i) *Subsistence agriculture*: developed due to the need to produce and consume fresh fruits and vegetables by households in most urban areas.

(ii) *High-technology agriculture*: located in urban fringe areas or large cities, capable of producing large agricultural output with high quality and ability to serve large markets, including some capacity for export. Examples of high-technology agriculture that provides for urban consumption needs and contributes select agricultural products for export can be found in Hanoi, HCMC, Da Nang, and Can Tho.

(iii) *Protective agriculture*: popular in industrial urban areas such as Bien Hoa, Viet Tri, and Dung Quat with a significant increase in tree cover for environmental protection.

(iv) *Agriculture for tourism*: transformed towards producing products to serve the needs of the tourism and hospitality sector (vegetables, flowers, ornamental plants, specialty aquaculture, etc.). This is prominent in cities such as Ha Long, Do Son, Da Lat, Nha Trang, Vung Tau that attract a lot of tourism.

(v) *Ecological agriculture*: this is a general trend for modern agricultural development that is environmentally friendly, aspiring to an ecological balance.

Having recently joined the group of middle-income countries, Vietnam remains focused on promoting economic growth while striving for synergies in technical, socio-economic, and ecological aspects of development. Policies such as Viet Nam's Socio-Economic Development Strategy (2021–2030) and Socio-Economic Development Plan (2021–2025) place emphasis on innovation, improved labour productivity, scientific and technological advances, improved quality of human resources, promotion of comparative advantages, and proactive international integration.

The Fourth Industrial Revolution, with breakthroughs in technology, especially digital technology, responds directly to the above aspirations and has spurred a growing trend for high-technology agriculture engaging both digital and analogue solutions to optimise the production system (Hosseinifarhangi et al., 2019). Digital technology solutions aimed at optimisation of inputs, and technology-aided monitoring of risks can be environmentally sustainable and yield superior productivity compared to traditional production methods (Ibidem).

Vietnam is among the 10 countries most impacted by climate change globally and agriculture constitutes its most vulnerable sector (UNDP, 2022). Solutions that contribute to resilience, adaptation, sustained and growing productivity, and quality of agricultural products are seen as crucial and actively sought for in efforts to meet SDG2 Targets. In fact, in the last 5 years, science and technology have contributed about 30–35% to the growth value of agriculture in Vietnam (Ngoc, 2023). The adoption of high-technology solutions informed by scientific research is, therefore, considered a necessary step to ensure food security.

2.2 Lessons Learned and Motivations for Expanding Urban Agriculture and High-technology Farming in HCMC

A key issue for urban food security in HCMC is the high in-migration and the attendant need for housing and infrastructure, which reduce the area of agricultural land while simultaneously growing demand. According to statistics from the United Nations Population Fund (UNFPA), rural migrants presently account for a third of the population of HMCM, and there is no end in sight for this trend (Government of the Socialist Republic of Vietnam, 2011). Between 2010 and 2015, productive agricultural area decreased by an average of 700 ha per year, and this accelerated to 1,000 ha per year by 2020 (Department of Agriculture and Rural Development of HCMC, 2020; HCMC People's Committee, 2021b). Presently, HCMC's agricultural production capacity falls significantly below demand (HCMC People's Committee, 2022). Even its most productive sector of vegetables and flowers is only able to supply 28% of demand (Table 2.1; Xuan, 2021). In addition to the reduced agricultural land fund, fragmentation of the current production scale and the prevalence of traditional farming practices contribute to this. The area presently under so-called high-technology farming accounts only for about 0.44% (500 ha) of the total agricultural land area (Department of Agriculture and Rural Development of HCMC, 2022). There is an evident need for consolidation of production areas and application of high-technology solutions that support large-scale consistent production capacity and access to trained workers. In addition, there is a need to deal with epidemics and crop losses associated with climate change impacts.

Table 2.1. Ability of Urban Agriculture to Meet Demand in HCMC.

Sector	Capacity to Meet Demand (%)
Vegetables	28%
Live pigs	11%
Live cattle	19.7%
Seafood	14%
Poultry	1.2%

Source: Data from Xuan (2021).

Accordingly, by 2030, the city aims to increase the supply capacity of all types of food by 15% from 2020 (Department of Agriculture and Rural Development of HCMC, 2021). By 2045, agriculture should constitute a modern economic sector, characterised by technologically advanced processes at every stage of the supply chain from production, processing, and preservation to market connectivity (Department of Agriculture and Rural Development of HCMC, 2022). But realising this aspiration requires a change in thinking in agricultural production, with transformation of the crop-animal structure in accordance with the circular economy model to promote local food and urban food self-sufficiency. There are signs that this change is already beginning as the rapid urbanisation and climate change impacts are causing a reluctance to invest in traditional agricultural production. Moreover, the young, qualified workforce is attracted to the industry and services sectors, leading to increased labour shortages in agriculture. This provides a push for new solutions to develop urban agriculture to adapt to climate change and create jobs and a source of stable income for urban residents.

Research evidences high willingness to adopt technological solutions among urban food growers. A survey of urban growers in HCMC by To (2023) found that 97.63% of agricultural households agreed that biotechnology must be applied in breeding and seed selection, 96.83% of households supported a stronger focus on farming and animal husbandry techniques, and 97.10% of households agreed to promote the application of high-technology machinery in production in general. Besides, 95.25% of surveyed urban agricultural businesses requested the application of technology in work management, and 93.67% expressed the need for technological improvements in harvesting and post-harvest preservation (Vu, 2022).

There is good potential in meeting SDG Target 2.3: increase labour productivity and income of agricultural sectors. An increase in high-technology production that is able to grow productivity and counter the loss of agricultural land can also help address Target 2.1: access to safe and nutritious food for all. But the shift towards high-technology solutions must not happen at the expense of food safety. To this end, the HCMC Food Safety Management Board was established in 2016, with the goal of establishing a uniform process for food safety management, improving coordination mechanisms, and enhancing the effectiveness of food safety standards through regular inspections and fines for non-compliance (Food Safety Management Board of Ho Chi Minh City, 2020; Government, 2017).

3. COMMUNITY AWARENESS AND CRITERIA FOR SUSTAINABLE URBAN AGRICULTURE MODELS FOR THE CONTEXT OF HCMC

This section reports on research findings from the urban resilience from agriculture project by Van Lang University. A key aim of the project was to engage university and school teachers and students, and local farmers and farming businesses to address gaps in awareness, knowledge and skills pertaining to urban agriculture within communities (Gulyas & Edmondson, 2021) and to develop workable urban agriculture models for HCMC.

During the project, surveys, interviews, and workshops were undertaken with residents and growers across both central and suburban districts of HCMC to gauge awareness of urban agriculture, uptake and attitudes towards high-technology solutions in agricultural production, and needs and expectations of growers and residents for a suitable urban agriculture model in the city (Table 2.2). Additionally, a trial greenhouse was constructed on the university campus, containing a vertical agriculture model with a

Table 2.2. Urban Residents' Perceptions Regarding Urban Gardens.

Summary of Questions/Answers	Percentage of Households Answering that Agree with the Statement of the Given Question (%)			
	2020	2022	2023	
Gardening helps connect people with nature and helps family members become closer	65	89	93	
The vertical street garden model (wall garden) is suitable for most housing types in urban areas	71	96	95	
Urban gardening should apply automatic irrigation technology to help reduce labour and save irrigation water	89	92	93	
Gardening helps improve health and connect neighbours emotionally (thanks to exchanging vegetables and fresh food)	60	87	91	
Gardening helps beautify the house, increase real estate value, and minimise the negative effects of climate change	80	98	100	
Urban residents have easy access to technology and gardening techniques thanks to the media and the attention of educational institutions	50	89	91	
Vertical gardens and walled garden technology have been known and applied by urban people	27	69	90	

Source: Authors.

Table 2.3. Biomass Productivity and CO_2 Absorption Capacity of Produce Grown in the Trial Greenhouse.

Species	Vegetable Biomass (kg/m²)	CO_2 Absorption Capacity of the Plant (kg/m²)
Green mustard	27.9–30.4	48.0–52.4
Bok choy	33.0–34.2	56.7–59.0
Brassica juncea	30.1–35.4	51.9–61.1
Butterhead lettuce	22.3–22.7	38.47–39.1
Purple lettuce	21.4–23.6	36.8–40.7

Source: Authors.

range of vegetable crops grown over a two-year duration. The aim was to test productivity and sustainability criteria of selected crops across a range of models (Table 2.3).

To further improve awareness and acceptance of automated urban agriculture, the project included an awareness raising and education programme through which academics from Van Lang University shared project findings with agricultural business managers and employees of various administrative bodies such as the Vietnam Fatherland Front Central Committee, environmental resources department, women's organisations, and training and education department. In early 2023, an education programme on vertical gardening and urban agriculture was also provided to thousands of students and teachers in a number of high and middle schools in the city and neighbouring localities. The following subsections set out the key findings from this research and knowledge transfer.

3.1 Community Awareness of Urban Agriculture and Implications for Expansion

Findings from the survey of 300 residents in HCMC demonstrate that city dwellers are increasingly concerned about food quality, their health, and their living environment. Surveys undertaken in the central Binh Thanh and Go Vap Districts in HCMC show that living in townhouses decorated with green plants creates a feeling of closeness to nature and is favoured by the majority of the surveyed urban residents. Here, 'urban garden' refers to a type of growing system which urban residents can take advantage of to grow vegetables and flowers in a range of possible locations within urban dwellings (see Section 4.2). Survey results also provide evidence that there is a high willingness among households to establish such urban gardens and that there is wide-spread willingness to apply technology to production. More than 90% of survey respondents expressed the wish to adopt urban gardens and more than 99% perceived the application of 'vertical garden technology' and 'soilless culture' as an inevitable trend. Table 2.2 elaborates on survey findings among residents.

The findings from the trial greenhouse yielded positive results for both the productivity and CO_2 absorption capacity of vertical gardens (Table 2.3, Vu, 2022). The tested model was designed to be suitable for urban gardens in city centres and for households with terraces, balconies, or skylights. In addition to providing fresh vegetables, this garden model not only improves the look of urban dwellings but also benefits the ecological environment by the carbon fixation ability of the fresh vegetable cover.

Findings from a survey of farming businesses (cooperatives, cooperative alliances, and individual farm enterprises) in suburban areas show that more than 90% of the surveyed businesses already apply technology to agricultural production, including large-scale high-tech agriculture systems. There is every reason to expect this investment in high-technology solutions in the urban fringe to expand as the Department of Agriculture and Rural Development has recently started collaborating with relevant urban districts to incentivise the restructuring of urban agriculture at local scale to address fragmentation and production capacity in the period 2022–2025 in line with the city's strategic vision for agriculture to 2050 (Department of Agriculture and Rural Development of HCMC, 2021; HCMC People's Committee, 2021c, 2021d).

3.2 Deriving Criteria for Viable Urban Agriculture Models

The research findings together with evidence of the current situation in regard to urban agricultural capacity, technologies, and demand were collated for discussion in workshops with experts, researchers, and business managers. The aim of the workshops was to identify and evaluate issues related to urban agricultural development based on three aspects: economics, society, and environment. The outcomes of the discussions were used to construct a set of criteria for a sustainable urban agriculture model fit for HCMC to align with the HCMC People's Committee's vision for the development of urban agriculture in the city to 2050 (HCMC People's Committee, 2021d). The following criteria were proposed:

(1) Ability to meet the consumption needs of urban residents for fresh food with better quality and price than products produced in other localities (green vegetables, meat, eggs, milk, fresh seafood, etc.).

(2) Suitability for urban architecture and includes green space, a modern and civilised urban and rural landscape, a reduction of greenhouse gas emissions, and activities associated with tourism.

(3) Maximisation of the effective use of space, land, labour, and products suitable for small- and medium-scale production, as well as the specific conditions of the region.

(4) Ability to contribute to preserving, restoring, and developing typical and indigenous plant and animal breeds and providing new high-quality varieties to the city market and other localities throughout the country.

(5) Alignment with a circular agricultural economic model (using raw materials, fuel, by-products, and organic waste on site to recycle products for environmentally friendly agriculture).

(6) Ability to create new sources of income, help stabilise and improve the quality of life, and satisfy the increasing enjoyment needs of modern, civilised urban and rural populations.

(7) Product creation model applying advanced and modern techniques, technology, and production processes, promoting the advantages and central role in science, technology, and market of large urban areas, creating competitiveness.

(8) Ability to create linkage and cooperation in production, processing and connecting consumption markets in the city, domestically and for export.

(9) High capacity for adaptation to climate change and crops and livestock diseases.

4. RECOMMENDATIONS FOR URBAN AGRICULTURE MODELS IN HCMC

Based on the findings from the urban resilience form agriculture – project, and what we learned from experiences from high-technology agricultural models from other countries (Israel, China, Japan, etc.), we propose two distinct types of urban agriculture for HCMC.

(1) In central regions (highly urbanised areas), small- and medium-scale farming models serving mainly local needs for vegetables, flowers, ornamental plants, ornamental fish, and pets should be applied.

(2) In suburban areas, depending on the conditions of terrain, land, space, and building architecture in each area, the following agricultural production models that can be applied at larger scale and serve both local and export demand appear appropriate: (i) growing vegetables, mushrooms, flowers, orchids, apricot trees, ornamental plants (ornamental leaves, bushes, etc.), and urban plants; (ii) growing fruit trees and perennial industrial trees; (iii) afforestation and agroforestry production associated with eco-tourism; (iv) raising high-quality beef cattle and breeding cattle; (v) high-yield dairy cow farming; (vi) farms raising high-quality pigs and breeding pigs; (vii) high-quality aquaculture model; (viii) ornamental fish farming; and (ix) clean salt production.

4.1. Introducing High-tech Agriculture in Inner City Areas

The agricultural land area of the inner-city districts of HCMC is modest and highly fragmented (2,526 ha, accounting for about 11% of the total agricultural area of the city; HCMC Statistics Department, 2019, 2023). Therefore, a feasible and advisable solution would be to supplement the existing mostly traditional urban agriculture with vertical gardens (Figs. 2.1 and 2.2a and b) to make the most of the available space. Vertical growing systems are suited for growing most leafy vegetables, spices, and medicinal herbs to meet the growing demand for safe vegetables in the city, especially

Fig. 2.1.   Vegetable Plants Grown on the Terrace/Rooftop Garden.
Source: Authors.

premium vegetables and high-quality vegetables for restaurants, hotels, and other hospitality companies in the city. A broad adoption of vertical gardens in inner city areas could make a notable contribution to reaching the SDG2 targets that Vietnamese cities are presently struggling to meet.

4.2. Existing and New Production Models in Suburban Areas

4.2.1. Vegetable Farming

Large-scale vegetable production is concentrated mainly in the three suburban districts of Cu Chi, Hoc Mon, and Binh Chanh and accounts for about 89% (18,624 ha) of the entire city's vegetable growing area (HCMC Statistics Department, 2019, 2023). This type of vegetable production has good potential to contribute significantly to urban food security and the achievement of several

Fig. 2.2a.   High-technology Supported Urban Agriculture in the Suburbs.

Source: Authors.

SDG2 targets. The three suburban districts are able to accommodate a concentrated system of large-scale vegetable production in greenhouses or open air, depending on the plant species and output quality standards for each type of product. The common factor of these different production models should be the application of smart irrigation systems (Fig. 2.2a) and precision control of input materials (seeds, fertilisers, and pesticides) in accordance with good agricultural practice, organic production standards, good manufacturing practice, or equivalent, depending on the registration of the relevant cooperative or farming business. All enterprises participating in this large-scale production model should apply advanced technological solutions at every stage of the supply chain from production, processing, and preservation to market connectivity and should function in a coordinated manner to ensure stability of supply.

Fig. 2.2b.   Traditional Urban Agriculture in the Suburbs.
Source: Authors.

4.2.2. Production of Flowers, Ornamental Plants, and Urban Trees

While this type of urban agriculture does not have relevance for urban food security, it has good potential to enhance agricultural income (SDG Target 2.2). Tree cover and the production of flowers and ornamental plants also align with the HCMC's efforts to improve the urban environment, provide ecological benefits (for biodiversity and climate mitigation and adaptation), and improve the attractiveness of the city to tourists. Therefore, taking advantage of land and market potential and scientific and technological potential, especially in biotechnology, to promote the production of flower varieties (orchids Fig. 2.3, apricot flowers), ornamental plants, and urban trees is highly advisable. This is a sector that can achieve significant economic value from meeting the demand from the city's urban landscape development as well as export markets. Presently, the production area of flowers and ornamental plants is concentrated in Cu Chi, Binh Chanh, and Hoc Mon and is estimated to cover about 2,250 ha (HCMC Statistics Department, 2019, 2023). Of this, orchid production accounts for 300 ha, ornamental plants for 850 ha, and apricot and other urban trees for 550 ha each. Much of the orchid production already relies on high-technology solutions in breeding and production and has brought in an average revenue of $800 million to $1.2 billion per year (Government, 2018).

Fig. 2.3. Orchid Farming.
Source: Authors.

4.2.3. Fruit Trees and Long-term Industrial Trees

HCMC has well-established fruit tree production spanning approximately 8,000 ha (HCMC Statistics Department, 2019, 2023). The key tree species are rambutan, longan, mangosteen, and durian, concentrated along the Saigon River in Cu Chi district and Thu Duc city. Mangoes are concentrated in Giong Cat in Can Gio district on alluvial soils along rivers and canals. Citrus trees (grapefruit, lemon, orange, tangerine), guava, Siamese coconut are concentrated in the alum soil area in Binh Chanh district. An example of the application of high-technology solutions in this sector is the development of tissue culture bananas concentrated in large-scale planting areas in Cu Chi district (HCMC Party Committee, 2020). Developing and maintaining this type of agricultural production support the food security, income, and environmental targets of SDG2.

4.2.4. Models of Afforestation and Combined Agro-Forestry Production Associated with Eco-tourism

HCMC is determined to maintain the existing 33,372.44 ha of woodland as this supports the city's aspiration of functioning as an eco-tourism centre for the country (HCMC Statistics Department, 2019, 2023). Due to the varied landscape of the city integrating a river system, there is scope for industrial timber production, agro-forestry, and aquatic production such as fisheries. Digitally aided aquatic production has good potential to aid progress in access to safe and nutritious food for all (SDG2.1) and increasing income for farmers (SDG2.3) and can also contribute to the establishment of a circular agricultural economy in urban areas.

4.2.5. Animal Husbandry

HCMC is home to the Dairy Cow Company and industrial-scale pig and chicken companies (HCMC Statistics Department, 2019, 2023). Although it is not a particular strength, the livestock sector is nevertheless important for all of the SDG2 targets in urban areas. The production of high-quality animals, meat, eggs, and milk should be coupled with the establishment of local slaughterhouses and the provision of insurance cover to ensure biosecurity and food hygiene. Here, international collaborations should support the gradual modernisation of production technologies and the supply chain. HCMC should maintain its role as a centre providing high-yield dairy and beef cattle breeds and high-quality pork breeds. It should develop poultry herds on an industrial scale, especially egg-laying chickens and nurture cooperation across all sectors of livestock production to ensure safe and stable supply of fresh meat, eggs, and dairy products.

5. CONCLUSIONS

This chapter evidences the broad scope of urban agriculture present in HCMC and the city's strong position in applying high-technology solutions to urban agricultural production. With the potential vested in the city's natural conditions and human resources, along with favourable credit policies, access to scientific research, and international trade connections, we are confident that the city will be able to further develop urban agriculture that meets the need for nutrition among the growing population. Adopting and enhancing the range of high-technology supported urban agriculture models can improve food self-sufficiency and food security, increasing access to fresh foods among poor and marginalised households and augmenting agricultural productivity and incomes, thereby responding to the targets of SDG2 that Vietnam is still struggling to meet. Moreover, nurturing existing non-food-related agricultural production is not to be overlooked as it has important potential to enhance and diversify agricultural income and aligns with the city's aspirations pertaining to urban greening and eco-tourism.

The urban resilience from agriculture project demonstrated the scope for collaboration between growers, schools, and universities in furthering the adoption of high-technology agricultural solutions in the urban context. Research findings and stakeholder perspectives were brought together in workshops to derive criteria for viable and sustainable production models to align with existing policy goals. The criteria emphasise the importance of considering local conditions, both the nature and architecture of the area when expanding production. In the urban centre, vertical garden solutions can be applied for either small-scale subsistence production or a larger scale to support access to safe and nutritious food for all (SDG Targets 2.1 and 2.2) and to serve the needs of the hospitality sector. In suburban areas, larger scale high-technology reliant models have the potential to raise both productivity and income (Target 2.3) and increase sustainable food production that helps maintain ecosystems and enhances resilience to climate change and other disasters (Target 2.4).

To sum up, agricultural production models should be encouraged to develop according to the principle of circular agricultural economy, applying appropriate types of technology to make the most of urban space as well as creating safe and high-quality agricultural products, ensuring food security, and positively impacting the ecological and socio-economic environment. Van Lang University, in collaboration with local growers, has demonstrated that there is good scope for augmenting the uptake of high-technology solutions in all agricultural production sectors in HCMC and that the general attitude towards urban agriculture among residents is positive.

ACKNOWLEDGEMENTS

This work was supported by the British Council's Going Global Partnerships grant from the call for UK-Viet Nam Partnerships for Quality and Internationalisation (Grant ID GGPVN 3.6) led by

Middlesex University (UK) and Van Lang University (VN). Quyen Vu Thi would like to thank Van Lang University, Vietnam, for funding this work and to give many thanks to my lovely students: Le Minh Hieu, Nguyen Vu Huong Giang, Le Hoang Khoi, and Doan Huyen Trang for help with surveys for urban residents and for taking care of plants in the trial greenhouse.

REFERENCES

Department of Agriculture and Rural Development of HCMC. (2020, December 17). *The implementation of the program to develop high-tech agricultural applications in Ho Chi Minh City during the 2016–2020 period* (BC No. 271/BC-SNN). http://www.sonongnghiep.hochiminhcity.gov.vn/

Department of Agriculture and Rural Development of HCMC. (2021). Ability of urban agriculture to meet demand in Ho Chi Minh city. https://www.vietnamplus.vn/giai-quyet-diem-nghen-cung-ung-nong-san-thuc-pham-cho-tphcm-post737733.vnp

Department of Agriculture and Rural Development of HCMC. (2022, December 23). *Summary of agriculture and rural development sector in 2022 and plan implementation 2023* (Report No. 282/BC-SNN). http://ccptnt.vn/tong-ket-nganh-nong-nghiep-va-phat-trien-nong-thon-nam-2022-va-trien-khai-ke-hoach-nam-2023.html

Food Safety Management Board of Ho Chi Minh City. (2020, April 1). *Extending the pilot operation of the Ho Chi Minh City Food Safety Management Board* (Decision No. 446/QD-TTg). https://vpub.hochiminhcity.gov.vn/portal/VanBan/2020-8/kientoanbcdliennganh finalthang82020-210820150727.doc

Government. (2017, May 18). Situation of implementing policies and laws on food safety management in the period 2011–2016 (Report No. 211/BC-CP). https://vanban.chinhphu.vn/bao-cao-cua-chinh-phu-nam-2017/bao-cao-tinh-hinh-thuc-thi-chinh-sach-phap-luat-ve-quan-ly-an-toan-thuc-pham-giai-doan-2011-2016-10058633

Government. (2018, July 5). *The government on policies to encourage the development of cooperation and association in the production and consumption of agricultural products* (Decree No. 98/2018/ND-CP). https://vanban.chinhphu.vn/default.aspx?pageid=27160&docid=194092

Government of the Socialist Republic of Vietnam. (2011, February 28). Urbanization and the problems that arise. *Government Online Newspaper.* https://baochinhphu.vn/do-thi-hoa-va-cac-van-de-phat-sinh-10274542.htm

Gulyas, B. Z., & Edmondson, J. L. (2021). Increasing city resilience through urban agriculture: Challenges and solutions in the global north. *Sustainability, 13*(3), 1465. https://doi.org/10.3390/su13031465

HCMC Party Committee. (2020). *Socio-economic development strategy for 10 years 2021–2030* (11th Ho Chi Minh City Party Committee Resolution – Term 2020–2025). https://www.hcmcpv.org.vn/tin-tuc/nghi-quyet-dai-hoi-dai-bieu-dang-bo-thanh-pho-ho-chi-minh-lan-thu-xi-nhiem-ky-2020-2025-1491870713

HCMC People's Committee. (2021a, August 21). *The plan to ensure the supply of essential goods, food, and foodstuffs during the city's implementation period Urgent measures to prevent and control the Covid-19 epidemic according.* (Document No. 2798/KH-UBND; Resolution No. 86/NQ-CP dated August 6, 2021 of the Government). https://thuvienphapluat.vn/van-ban/The-thao-Y-te/Ke-hoach-2798-KH-UBND-2021-dam-bao-cung-ung-luong-thuc-thuc-pham-thiet-yeu-Ho-Chi-Minh-485522.aspx

HCMC People's Committee. (2021b, June 10). *The City People's Committee on approving the program to develop plant varieties, animals and high-tech agriculture in Ho Chi Minh City in the period 2020–2030* (Decision No. 2092/QD-UBND). http://congbao.hochiminhcity.gov.vn/cong-bao/van-ban/quyet-dinh/so/2092-qd-ubnd/ngay/10-06-2021/noi-dung/44429/44438

HCMC People's Committee. (2021c, November 25). *The City People's Committee on plan to implement the program to develop plant varieties, animals and high-tech agriculture in Ho Chi Minh City during the period 2021–2025* (Decision No. 3931/KH-UBND). https://chicucttbvtvhcm.gov.vn/hoat-dong-don-vi/dinh-huong-trong-trien-khai-thuc-hien-chuong-trinh-phat-trien-giong-cay-con-va-nong-nghiep-cong-nghe-cao-tren-dia-ban-thanh-pho-ho-chi-minh-1456.html

HCMC People's Committee. (2021d). *Approving the task outline urban agriculture development program in Ho Chi Minh City in the period 2021–2030, vision to 2050* (Decision No. 4309/QD-UBND). http:// congbao.hochiminhcity.gov.vn/tin-tuc-tong-hop/uy-ban-nhan-dan-thanh-pho-ho-chi-minh-ban-hanh-quyet-%C4%91inh-so-4309-q%C4%91-ubnd-ve-phe-duyet-%C4%91e-cuong-nhiem-vu-%E2%80%9Cchuong-trinh-phat-trien-nong-nghiep-%C4%91o-thi-tren-%C4%91ia-ban-thanh-pho-ho-chi-minh-giai-%C4%91oan-2021-2030-tam-nhin-%C4%91en-nam-2050%E2%80%9D

HCMC People's Committee. (2022, August 31). *Implementation currently ensuring food security in the area until 2030* (Decision No. 2957/QD-UBND on implementing Action Program No. 12-CTr/YU dated January 13, 2022 of the Standing Committee of the City Party Committee). http://www.sonongnghiep.hochiminhcity.gov.vn/tonghop/lists/posts/post.aspx?Source=/tonghop&Category=Ch%C6%B0%C6%A1ng+tr%C3%ACnh+-+b%C3%A1o+c%C3%A1o&ItemID=806&Mode=1

HCMC Statistics Department. (2019). *Statistical Year Book of Ho Chi Minh City 2018*. HCMC General Publishing House. https://www.gso.gov.vn/wp-content/uploads/2019/10/Nien-giam-2018.pdf

HCMC Statistics Department. (2023). *Statistical Year Book of Ho Chi Minh City 2022*. HCMC General Publishing House. https://thongkehochiminh.gso.gov.vn/Niengiam/Niengiam

Hosseinifarhangi, M., Turvani, M., Van Der Valk, A., & Carsjens, G. (2019). Technology-driven transition in urban food production practices: A case study of Shanghai. *Sustainability, 11,* 6070.

Langemeyer, J., Madrid-Lopez, C., Mendoza Beltran, A., & Villalba Mendez, G. (2021). Urban agriculture – A necessary pathway towards urban resilience and global sustainability? *Landscape and Urban Planning, 210,* 104055. https://doi.org/10.1016/j.landurbplan.2021.104055

Ministry of Planning & Investment. (2021). *The socialist republic of Viet Nam (summary report) national report 2020 progress of five-year – Implementation of sustainable development goals.* https://www.undp.org/vietnam/publications/national-sdg-report

Ministry of Vietnam Politics. (2020). *Conclusion No. 81-KL/TW dated July 29, 2020 on ensuring national food security until 2030.* https://tulieuvankien.dangcongsan.vn/he-thong-van-ban/van-ban-cua-dang/ket-luan-so-81-kltw-ngay-2972020-cua-bo-chinh-tri-ve-bao-dam-an-ninh-luong-thuc-quoc-gia-den-nam-2030-8426

Ngoc, Q. (2023, April 25). Science and technology contribute 35% of the added value of the agricultural sector. Hanoimoi. https://hanoimoi.vn/khoa-hoc-cong-nghe-dong-gop-35-gia-tri-gia-tang-cua-nganh-nong-nghiep-439555.html

To, T. T. T. (2023). *Hi-tech agriculture development in Ho Chi Minh City* [Ph.D. thesis, major in agricultural economics]. Vietnam Academy of Agriculture.

UNDP. (2022). *Country programme document for the socialist republic of Viet Nam (2022–2026).* https://www.undp.org/vietnam/publications/country-programme-document-viet-nam-2022–2026

Vu, T. Q. (2022). *Report on the potential for applying technology in agricultural production and the possibility of developing a vertical garden system in Ho Chi Minh City.* Van Lang University.

Xuan, A. (2021, August 31). Solving the 'bottleneck' of agricultural and food supply for Ho Chi Minh City HCM. *VNA/Vietnam.* https://www.vietnamplus.vn/giai-quyet-diem-nghen-cung-ung-nong-san-thuc-pham-cho-tphcm-post737733.vnp

3

CONSEQUENCES OF HEAVY RAIN ON PRODUCTIVITY OF SUGAR COMPANIES IN PERU: A CASE STUDY TOWARDS ACHIEVING SDG2 ZERO HUNGER

Juan Diego Zamudio Padilla[a] and Constanza Flores Henríquez[b]

[a]Hiroshima University, Japan
[b]University of Canterbury, New Zealand

ABSTRACT

Enhancing agricultural productivity is imperative for sustainable food production and aligns with the objectives of SDG2 target 2.3. This target aims to achieve a twofold increase in agricultural productivity by doubling the volume of production per labour unit by 2030. Higher education institutions have actively engaged in leveraging information, technology, and promotional strategies to bolster agricultural productivity in developing nations, contributing to the broader SDG2 Zero Hunger agenda. However, limited research addresses the impact of heavy rainfall on productivity, utilising robust methodologies such as differences-in-differences and advocating for prospective causal investigations. This study specifically investigates the repercussions of heavy rain on Peruvian sugar companies, utilising data from the Lima Stock Exchange and the National Institute of Statistics for 18 companies spanning 2014 to 2017. Emphasising the negative consequences of heavy rain in 2017 attributed to the El Niño phenomenon, the research underscores regional variations, particularly noting heightened adversities for sugar companies in Northern Peru. This study not only elucidates the factors contributing to these negative consequences but also suggests diverse research approaches for comprehensive understanding and causal impact assessments, contributing significantly to interdisciplinary research in disaster management and impact economic evaluation. As an imperative step towards achieving Zero Hunger goals amid the prevailing climate emergency, the study concludes by presenting recommendations to mitigate the challenging effects of El Niño on the agroindustry.

Keywords: Productivity; El Niño phenomenon; higher education engagement; heavy rain consequences; causal impact evaluation; agriculture

INTRODUCTION

Higher education's substantive role in achieving the United Nations sustainable development goal 2 (SDG2) Zero Hunger, particularly SDG target 2.3 aiming to double agriculture pro-ductivity by increasing the volume of production per labour unit (United Nation, 2024), is crucial, especially in developing nations. Climate change's impact on agriculture in these regions has been studied extensively (Alam et al., 2023; Chen et al., 2023; Singh et al., 2023). However, there is limited research on the link between extreme climate change and productivity (Firmansyah & Nurjani, 2023; Lal & Singh, 2023), especially in South American agriculture with higher education involvement.

This section sets the context, providing an overview of hunger issues and our research agenda objectives. It highlights the role of higher education in SDG2 Zero Hunger, emphasising SDG2.3 target. The influence of phenomena like El Niño on agricultural productivity is explored, revealing the intricate connection between climate events and global food security (United Nations, 2023). The significance of productivity, particularly in SDG2.3, has gained prominence. This goal emphasises assessing production efficiency per labour unit, especially in the crucial agricultural sector, a primary source of sustenance globally. Unlike the narrow employment-centric view in previous goals, this nuanced metric has broader implications, crucial for addressing poverty for a substantial portion of the world's impoverished population, estimated at 80% (United Nations, 2015; World Bank, 2023).

Building on the global context of SDG2.3 target, from 2016 to 2020, the Universidad Nacional Mayor de San Marcos witnessed significant changes in its International Technical Cooperation Office, influenced by efforts led by the staff (United Nation, 2022; UNMSM, 2017). The Covid-19 pandemic in 2020 prompted a recalibration of efforts, leading some staff members to pursue academic opportunities abroad. The office's engagement in the Erasmus Plus Capacity Building programme in 2017 faced linguistic limitations, allowing only projects manageable by proficient staff to persist, administratively led by the first author of this chapter for around 5.8 million euros (ACACIA, 2019; FLAUC, 2017). The team prioritised contributing to education, university development, public education, and rural economic development, adopting a collaborative and research-oriented approach since 2016 (OGCRI, 2024; UNMSM, 2024; Zamudio, 2024a, 2024b).

Organising competitions, the team aimed to build community trust and encourage student participation, emphasising knowledge transfer. The head of the office expanded activities globally, collaborating with embassies from various countries. The team sought to exemplify development for universities, aiming to transform into a first-class, research-oriented institution. Their efforts extended beyond internal gains, emphasising collaboration with other countries and public universities. In Peru, diplomatic support led to the establishment of the Peruvian Network of Internationalization in 2018, involving 11 universities and supported by the Embassy of Japan. Collaborative initiatives were launched, resulting in an agreement between 12 universities in Peru and 12 in Japan, emphasising a cooperative approach towards SDGs (Embassy of Japan in Peru, 2024).

A collaborative initiative involving Universidad Nacional Mayor de San Marcos, University of Liverpool, and Xi'an Jiaotong-Liverpool University focused on a business challenge initiative related to climate change. This initiative, originating from collaborative efforts facilitated by the British Council Peru, Liverpool University, and Universidad Nacional Mayor de San Marcos, addressed SDGs, highlighting women in technology and the impact of climate change on rural infrastructure (Touchstone, 2021).

Dealing with developed countries presented administrative challenges initially, but concrete challenges proposed for students proved effective. San Marcos University collaborated with the Embassies of the United States and the United Kingdom, actively promoting SDGs in education. The university's engagement in education conferences positioned Peruvian public universities alongside research-oriented institutions.

In 2020, significant initiatives unfolded. Initially, participation in short mobility programmes at German and Polish universities, focusing on topics such as Doing Business in Latin America, Venture Capitals with a Peruvian perspective, and Global Healthcare Management (Zamudio, 2024a, 2024b), aimed to explore diverse methods and strategies. These courses centred on food security, climate change impact, economic evaluation, and global health legal and management issues, engaged students globally in identifying SDG-related gaps for innovative solutions.

Simultaneously, both authors undertook the sustainable development global leadership programme at Hiroshima University, sponsored by JICA. From 2020 to 2023, they shared a commitment to climate change research, adopting a microeconomic perspective and causal inference methods. This research approach, rooted in their experiences at Hiroshima University, focuses on evaluating El Niño's impact on SDG2 Zero Hunger targets, particularly emphasising SDG2.3 to double agricultural productivity by 2030 through increased production volume per labour unit.

This commitment to climate change research aligns seamlessly with their exploration of El Niño-Southern Oscillation (ENSO), a recurrent climate pattern driven by ocean–atmosphere interactions in the tropical Pacific. 'El Niño' refers to the warm phase of ENSO, while La Niña is the cold phase, occurring every two to seven years, with neutral phases in between. One of the consequences of ENSO is ocean temperature anomalies in South America, affecting local rainfall patterns with global economic implications, particularly in agriculture and fishing (Lal & Singh, 2023).

Since the impactful El Niño events in 1972/1973, 1982/1983, 1997/1998, and 2015/2016, ENSO has become crucial for Latin American economies. Peru, with coastal exposure, faces heightened vulnerability due to warmer ocean temperatures (Instituto Del Mar Del Perú, 2024; Sanabria et al., 2018). ENSO's consequences extend globally, affecting agricultural and fishing sectors critical for the global food supply chain (Yglesias-González et al., 2023). The record El Niño event in 2015–2016 impacted over 60 million people worldwide. Repercussions included severe hurricanes on the East coast of North America, wildfires in Western Canada, and sustained droughts in India and Pacific Island countries (Thomalla & Boyland, 2017).

Globally, El Niño and La Niña have varying impacts on agriculture. In China, La Niña has a marginal positive contribution to grain productivity, while El Niño has a marginal negative impact. In southern Ethiopia, differing effects on crop productivity show resilience to El Niño's impact on rainfall distribution. South America, especially coastal deserts, experiences unique effects, with positive short-term impacts on green growth during extreme El Niño events (Haile et al., 2021; Li et al., 2020; Vining et al., 2022). Thus, El Niño's consequences are intricate, spanning sectors and regions globally. A comprehensive understanding of its dynamics is crucial for a practical evaluation of its impact on agriculture productivity (Haile et al., 2021; Instituto Del Mar Del Perú, 2024; Li et al., 2020; NOAA, 2024; Sanabria et al., 2018; Thomalla & Boyland, 2017; Vining et al., 2022; Yglesias-González et al., 2023).

BACKGROUND OF PRODUCTIVITY DYNAMICS IN THE PERUVIAN SUGAR INDUSTRY

The global sugar industry has undergone significant transformations, with a particular focus on the dynamic balance between production and consumption. The International Sugar Organization (2017) reported a deficit in the global balance between 2015 and 2016, triggering shifts in production patterns and altering consumption trends. Notably, the industry faced heightened challenges during the extreme El Niño event of 2015/2016, which impacted daily sugar prices and influenced international trade dynamics (Index Mundi, 2024). This study focuses on a critical case of sugar productivity within this evolving landscape.

During the critical period of 2015/2016 affected by El Niño in the volume of production, the global sugar industry grappled with a deficit of 5.2 million tonnes, a trend that persisted and intensified in 2016/2017, reaching 5.9 million tonnes (International Sugar Organization, 2017). Concurrently, the International Sugar Agreement price reveals significant fluctuations in daily sugar prices. The lowest point, at 0.25 USD per kg of sugar in 2014/2015, sharply contrasted with the highest point of 0.49 USD per kg in 2015/2016 (Index Mundi, 2024). These shifts prompted numerous countries to reconsider their international trade strategies within the sugar market. Thus, the sugar industry represents a global industry to be attended to for its repercussions in the world under extreme climate change, and its productivity needs to be analysed.

El Niño's impact on the Andean countries, led by Peru, showcased mixed effects during the 1982–1983 and 1997–1998 events. While the economic challenges were notable, increased rainfall in Northern Peru's dry areas had positive effects, reducing poverty rates by 5% in communities reliant on dry forests (Pécastaing & Chávez, 2020). However, comprehending the diverse impacts of El Niño on poverty and hunger in developing nations remains a challenge (Karuniasa & Pambudi, 2022). Studying Peru's experience with El Niño is crucial for understanding its consequences on productivity and addressing the mixed effects on the country's landscape.

As the world pursues SDG2, Zero Hunger, this study evaluates productivity in Peruvian sugar companies, emphasising SDG2.3 target to double agricultural productivity increasing the volume of production per labour unit by 2030. Recognising their role in global food provision, the research employs causal methods to assess the impact of ENSO and acknowledges the effects of extreme climate events on their performance. The study delves into the dynamics of the Peruvian sugar industry in the context of global shifts and El Niño, highlighting the importance of SDG2.3 and the involvement of higher education in achieving global food security and doubling production per labour unit.

DESIGN OF THE DIFFERENCE-IN-DIFFERENCES METHOD

To understand the impact of heavy rain on sugar companies, our study employs the difference-in-differences methodology. This method relies on two critical conditions: the existence of a parallel trend before the heavy rain and a distinct deviation in the treatment group compared to the control group (Bertrand et al., 2004; Columbia University, 2024; Roth, 2022). Our statistical examination reveals a parallel trend before the onset of heavy rain and a substantial deviation during periods of El Niño-induced heavy rainfall on the North Coast of Peru.

Several notable factors contribute to the strength of our analysis. Firstly, the entities under scrutiny are homogeneous, comprising companies within the same sugar industry. Furthermore, they share a common listing on the Lima Stock Exchange and collectively represent significant players in Peru's sugar landscape.

Building upon this contextual foundation, the temporal alignment of our study with a climatic event of significance enhances the relevance and specificity of our investigation, between 17 March 2017 and 23 March 2017, heavy rain significantly impacted the North Coast of Peru, affecting provinces like Piura and Lambayeque, leading to increased sugarcane growth. However, this rain resulted in low saccharose levels, a key component for sugar production, negatively impacting overall output (Pierce, 2017). La Libertad province also faced from during January to March 2017.

While heavy rain events are not new, the 2017 incident raised concerns, especially for companies, causing significant damage (Bayer et al., 2014). The scarcity-induced rise in sugar prices triggered an investigation by the Peruvian Institution for Consumer Protection, exploring potential market collusion (La República, 2017). The heavy rain event also caused a surge in prices, potentially distorting productivity analysis by inflating sales. To focus solely on the quantity of production generated by labour units in Peru's sugar industry, we intentionally eliminate the price effect. In alignment on this objective, our study employs two distinct groups: the treatment group, which consists of companies impacted by heavy rain on the North Coast of Peru, and the control group, which consists of companies not affected by heavy rain in the rest of Peru.

DATA

In this study, we utilised the Lima Stock Exchange to access information on 18 key agro-industry companies listed in Peru, focusing on their significance in the agricultural sector. The Lima Stock Exchange, established in 1857, provided comprehensive company details, historical prices, financial information, and corporate memories (Bolsa de Valores de Lima, 2024). Our deliberate choice of the Lima Stock Exchange, as opposed to alternatives like the Stock Market Superintendence, was influenced by considerations of information provider reliability and data organisation by economic sectors. Insiders' possession of non-public information can lead to asymmetric information in financial markets, influencing their behaviour (Flannery, 1986). The decision to list companies is pivotal for market transparency and investor decision-making (Ahmad et al., 2023).

To compile our company list (Table 3.1), we verified details through the National Superintendency of Customs and Tax Administration (SUNAT, 2024), ensuring accurate localisation and property verification. Exclusion criteria removed 12 companies, leaving 6 dedicated to the sugar industry. Our objective is to analyse the volume of sugar production per labour unit for each company by aggregating yearly sugar sales and labour force data.

We adopted tonnes of sugar sales as a production variable, simplifying productivity calculations, given the sugar price increase and complexities during 2017's extreme El Niño (Ranganathan & Benson, 2020). Despite diversified products, the primary focus of the six companies was sugar. Table 3.2 presents the productivity calculation, assuming all labourers are exclusively dedicated to sugar production. The result is in tonnes per labour unit.

Table 3.1. Sugar Company Description, Location, and Economic Activity.

Company	Location		Economic Activity	
	Peru	Region-Province	Product	Economic Group
AGROINDUSTRIAS SAN JACINTO SOCIEDAD ANONIMA ABIERTA	North	La Libertad-Ascope	White and blonde sugar	Gloria
CARTAVIO SOCIEDAD ANONIMA ABIERTA (CARTAVIO S.A.A.)	North	La Libertad-Ascope	White and blonde sugar	Gloria
CASA GRANDE SOCIEDAD ANONIMA ABIERTA	North	La Libertad-Ascope	White and blonde sugar	Gloria
AGROINDUSTRIAL LAREDO S.A.A.	North Centre	La Libertad-Trujillo	White sugar	Gloria
AGROINDUSTRIAL PARAMONGA S.A.A.	Centre	Lima-Barranca	Blonde sugar	Wong
CENTRAL AZUCARERA CHUCARAPI PAMPA BLANCA S.A.	South	Arequipa-Islay	White sugar	Various
Companies (N = 6)				

Note: The author succinctly summarises crucial information regarding the location[a] and economic activities[b] of the units under analysis, utilising data from Bolsa de Valores de Lima (2024).
[a]The geographical focus aims to identify proximity to the North Coast of Peru, a region prone to heavy rainfall, offering insights into potential productivity consequences on the analysed units.
[b]The emphasis on economic activities is pivotal, confirming the alignment of all units within the same industry and their primary production of identical or closely related products, ensuring homogeneity for meaningful comparisons.

Table 3.2.   Productivity – Produced Tonnes (Bag of 60 kg) of Sugar per Labour Unit.

Company	Years			
	2014	2015	2016	2017
AGROINDUSTRIAS SAN JACINTO SOCIEDAD ANONIMA ABIERTA	75	79	87	74
CARTAVIO SOCIEDAD ANONIMA ABIERTA (CARTAVIO S.A.A.)	151	138	148	107
CASA GRANDE SOCIEDAD ANONIMA ABIERTA	108	90	83	56
AGROINDUSTRIAL LAREDO S.A.A.	101	90	80	85
AGROINDUSTRIAL PARAMONGA S.A.A.	103	101	99	102
CENTRAL AZUCARERA CHUCARAPI PAMPA BLANCA S.A.	20	12	8	10

Note: The author calculates sugar production in tonnes (60 kg bags) per labour unit for each company from 2014 to 2017 using data from Bolsa de Valores de Lima (2024). This calculation involves dividing the production sold each year by the number of labourers[a], serving as a metric for agriculture productivity across all the companies.

[a]The detailed calculation method explained in the data section provides a specific example for gauging the volume of production per labour unit, supporting the objective of doubling agriculture productivity outlined in SDG2.3.1, with the sugar case as an illustrative reference.

ESTIMATION

Our quantitative objective was to analyse the impact of heavy rain on the productivity of sugar companies in Peru. We did this by estimating the Average Treatment Effects on the Treated over the control group. Following the work of Pécastaing and Chávez (2020), we empirically design the following econometric model.

$$P_{ij} = \theta + \beta + \gamma + \delta \text{Company}_i * \text{Heavy Rain}_t + \varepsilon \qquad (1)$$

In Equation (1), the variable P_{ij} denotes productivity, measured as the sugar tonnes sold per labour (in 60 kg bags). This metric serves as the quantitative outcome under investigation in this study. It is contingent upon the interaction term δCompany$_i$* Heavy Rain$_t$, where δ represents the parameter of interest. The information is aggregated at the company level, with (i) denoting the company as the primary unit of observation and (t) signifying the occurrence of heavy rain in the year 2017. The fixed effect for the year is denoted by β, while γ represents the fixed effect at the company level. The error term ε accounts for the normal standard error associated with the annual productivity level of the company. Finally, θ is the constant term in this equation.

RESULTS

Within the scope of SDG2.3, aiming to double agricultural productivity and production volume per labour unit by 2030 (United Nation, 2024), the 2017 heavy rain (Pierce, 2017; Sanabria et al., 2018) significantly impacted the sugar industry on the North Coast of Peru. This setback jeopardised the industry's goal aligned with SDG2 Zero Hunger and SDG2.3, leading to a reduction in sugar volume production per labour unit. The ensuing challenges in meeting sugar demand resulted in increased prices, hindering progress towards SDG target 2.3.

Fig. 3.1, created using Google Sheets (2024), presents a comparative analysis of productivity. It highlights the contrast between the treatment group (Gloria Economic Group in the North) and the control group (companies in Lima and Arequipa). The control group faced negative productivity consequences, particularly in 2015, 2016, and 2017. Conversely, the northern group exhibited resilience in 2016 but experienced a decline in productivity in 2017.

Table 3.3, analysed using R version 4.3.2 (2024), supports earlier studies on heavy rain's adverse impact on company productivity. While lacking statistical significance, our findings align with Karuniasa and Pambudi (2022) observations in Indonesia. Pécastaing and Chávez (2020) reported a 5% lower likelihood of households facing poverty in dry forest areas, emphasising the role of dry

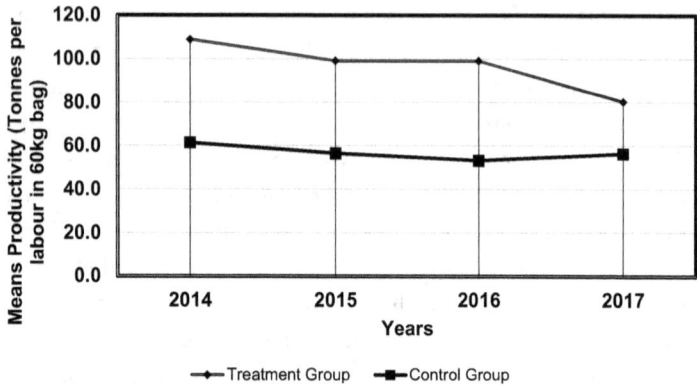

Fig. 3.1.   Treatment and Control Group Productivity Means from 2014 to 2017 of Sugar Companies in Peru.

Source: The author uses Excel to calculate mean productivity (tonnes per labour in a 60-kg bag)[a] from 2014 to 2017, presented in a figure. Two groups, a treatment, and a control group[b], show a parallel trend pre-2017, coinciding with heavy rainfall. The application of the difference-in-differences methodology leverages this trend for robust comparative analysis. In 2017, heavy rain adversely affected the treatment group, resulting in a more pronounced drop in mean productivity (tonnes per labour in a 60-kg bag) compared to the control group during the same period.
[a]Measured as tonnes of sugar per labour in a 60-kg bag, aligning with the outcome indicator for SDG2.3.1, doubling agricultural productivity, by duplicating the volume of production per labour unit.
[b]Detailed descriptions of the treatment and control groups are available in the difference-in-differences method section, specifying the treated group's location on the North Coast of Peru and the control group elsewhere.

forests in mitigating flood consequences. Including the 2017 heavy rain impact in future studies is crucial for accurate assessments.

Maintaining methodological rigour, our research suggests the importance of expanding units, extending analysis periods, utilising high-frequency databases, and focusing on precisely defined treatment areas to achieve statistically significant results. This approach contributes to a nuanced understanding of heavy rain events' dynamics on productivity, especially in the context of sugar production in Peru. Building on this impact evaluation, the reverberations of productivity challenges induced by heavy rain extend beyond corporate entities. Productivity challenges induced by heavy rain extend beyond corporate entities to impact higher education institutions, exemplified at the National University of Trujillo, where faculties grappled with disruptions (Noticias Trujillo, 2017).

Table 3.3.   Difference-in-Differences Estimation Model Results.

	Dependent Variable
	Productivity
Treated	43.906**
	(16.799)
Year	−3.202
	(8.400)
did	−15.617
	(26.562)
Constant	6,510.929
	(16,929.250)
Observations	24
R^2	0.286
Adjusted R^2	0.179
Residual std. error	35.636 (def. = 20)
F Statistic	2.670* (df = 3; 20)
Note:	* $p < 0.1$; ** $p < 0.05$; *** $p < 0.01$

Note: The author uses R software version 4.3.2 to calculate the difference-in-different model of sugar productivity, focusing on sold tonnes produced per labour unit and basing the estimation on an empirical econometric equation[a]. In the Bolsa de Valores de Lima (2024) database, the treatment year is set as 2017 due to heavy rain in the North Coast of Peru, assigning a value of 1 for that year and 0 for 2014 to 2016. For the treated group, a dummy variable is defined as 1 for companies on the North Coast of Peru and 0 otherwise. The dummy for the difference-in-difference is derived from the product of the dummy for treated years and the dummy for the treated area. The results of the equation applied to a balanced data panel with 24 observations yield a negative difference-in-differences estimator that is not statistically significant[b].
[a]The econometric equation is detailed in the estimation section.
[b]This result is expounded upon and compared with the previous findings of a negative consequence of heavy rain on agriculture productivity.

Rotas and Cahapay (2020) studied heavy rain's impact on undergraduate students in the Philippines, revealing limitations in class attendance and knowledge acquisition, particularly during adverse weather.

Heavy rain also disrupts high schools, with 1,523 schools in Peru requiring reconstruction due to adverse weather conditions during 2015/2017 (Ojo Público, 2020). This impediment to class attendance aligns with studies on high-school students in rural India and Ethiopia, where heavy rain deters attendance, affecting human capital acquisition and hindering the achievement of productivity targets (Falaris, 2023; Shah & Steinberg, 2017). These challenges have far-reaching implications for the Peruvian economy, contributing to a slowdown with only 2.5% growth in 2017, down from 4% the previous year (Central Bank of Peru, 2017). This intricate interplay underscores the connection between environmental challenges, education, and economic outcomes within the broader context of SDG2.3 towards achieving double agriculture productivity.

RECOMMENDATIONS

To comprehensively assess the impact of extreme climate change on companies and households, Pécastaing and Chávez (2020) employed a causal approach with annual data, categorising households into two comparable groups. Using the Difference-in-differences method, they traced outcomes over time. However, to achieve a more nuanced and precise evaluation, incorporating monthly data and obtaining detailed climate measurements could enhance accuracy and provide a micro-level observation for impact assessment.

This methodological refinement aligns with the overarching goal of enhancing resilience in both large- and small-scale agricultural production, crucial for achieving SDG2.3 (United Nation, 2024). The collaborative efforts between higher education, companies, and organisations are essential for science-based policies and effective strategies. Resilience-building initiatives, especially for small-scale producers facing barriers to technology adaptation, can play a significant role in mitigating poverty and hunger (Azadi et al., 2023). Programmes designed and evaluated by higher education globally contribute to shaping science-based policies that address the challenges posed by extreme climate change in agriculture.

In the pursuit of SDG2 (Zero Hunger), fostering climate-resilient productivity becomes vital. Drawing insights from countries like Japan, students need to learn causal method evaluations for climate change impact assessment. By employing innovative evaluation methods, students can contribute to scientific literature and become leaders in promoting small-scale farmer productivity.

Their focus should encompass skill development, resilience enhancement, and understanding the complexities of extreme climate change.

Initiating short activities, like a dedicated summer programme on SDG2.3, can engage students, fostering collaboration and generating impactful interventions. Participating in analytical circles, gaining access to global research examples, and conducting primary data research in agriculture areas during visits to small farmers provide practical experiences for undergraduate students (Zamudio, 2024a, 2024b). Forming agreements between public universities in Peru and global institutions facilitates collaborative learning and experience-sharing. Prioritising research-oriented initiatives allows for focused interventions in climate-affected areas. Rigorous causal analyses measure the effectiveness of interventions, involving undergraduate students actively in increasing productivity using various strategies.

Ensuring policy effectiveness demands strategic planning. Government collaboration with higher education can address information barriers for companies of all scales, particularly during events like El Niño. Evaluating programmes for productivity enhancement and rural economic development, in addition to export-oriented agriculture, requires an open, cooperative approach. Motivating the government to open data access and collaborate with research centres is crucial. Organising open data access panels as primary raw data, labelling balanced panels, and developing collaborative platforms enhance research quality. Generating specific variables in databases, like measuring production per labour unit, ensures accurate and valuable insights through collaborative efforts between government entities and research institutions. The collective integration of these academic endeavours contributes to a comprehensive and impactful approach to addressing the challenges posed by extreme climate change on productivity and agricultural sustainability.

CONCLUSIONS

The study investigates the impact of heavy rainfall on agroindustry productivity, a key factor in achieving SDG2 Zero Hunger.

Focusing on doubling agricultural productivity by enhancing the volume of production per labour unit, the analysis employs observational rainfall and economic data from 2014 to 2017, particularly in the North Coast area of Peru 2017. Utilising data from six Lima Stock Exchange-listed companies during 2014–2017, with heavy rainfall as the treatment, our analysis reveals negative consequences on productivity in the North Coast. While statistically significant findings were limited due to data constraints, the study underscores broader societal impacts, including disruptions in higher education, reconstruction challenges, and economic slowdown.

Future research efforts should refine methodologies, and foster collaboration among academic institutions, industries, and policymakers, emphasising the crucial role of higher education in building resilience and contributing to SDG2 Zero Hunger and the target 2.3. The study advocates for comprehensive approaches, frequent data for precision, and collective responsibility through collaborations between higher education, government, and statistical institutions. Initiatives like summer programmes and interinstitutional agreements can further drive progress towards SDG2.3.

REFERENCES

ACACIA. (2019). *Erasmus plus capacity building project.* https://erasmus-plus.ec.europa.eu/projects/search/details/561754-EPP-1-2015-1-CO-EPPKA2-CBHE-JP

Ahmad, M. M., Hunjra, A. I., & Taskin, D. (2023). Do asymmetric information and leverage affect investment decisions? *The Quarterly Review of Economics and Finance, 87*, 337–345.

Alam, M., Ali, M. F., Kundra, S., Nabobo-Baba, U., & Alam, M. A. (2023). Climate change and health impacts in the South Pacific: A systematic review. In U. Chatterjee, A. O. Akanwa, S. Kumar, S. K. Singh, & A. Dutta Roy (Eds.), *Ecological Footprints of Climate Change: Adaptive Approaches and Sustainability* (pp. 731–747). Springer, Cham. https://doi.org/10.1007/978-3-031-15501-7_29

Azadi, H., Ghazali, S., Ghorbani, M., Tan, R., & Witlox, F. (2023). Contribution of small-scale farmers to global food security: A meta-analysis. *Journal of the Science of Food and Agriculture*, *103*(6), 2715–2726. https://doi.org/10.1002/jsfa.12207

Bayer, A. M., Danysh, H. E., Garvich, M., Gonzálvez, G., Checkley, W., Álvarez, M., & Gilman, R. H. (2014). An unforgettable event: A qualitative study of the 1997–98 El Niño in northern Peru. *Disasters*, *38*(2), 351–374.

Bertrand, M., Duflo, E., & Mullainathan, S. (2004). How much should we trust differences-in-differences estimates? *The Quarterly Journal of Economics*, *119*(1), 249–275. https://doi.org/10.1162/003355304772839588

Bolsa de Valores de Lima. (2024, February 15). *Empresas*. https://www.bvl.com.pe/

Central Bank of Peru. (2017). *Memory 2017*. https://www.bcrp.gob.pe/docs/Publicaciones/Memoria/2017/memoria-bcrp-2017.pdf

Chen, L., Xu, Z., Zeng, J., Zhu, G., Liu, X., & Huang, B. (2023). Sedimentary records of phytoplankton communities in Sanmen Bay in China: The impacts of ENSO events over the past two centuries. *Water*, *15*(7), 1255. https://doi.org/10.3390/w15071255

Columbia University. (2024, February 13). *Difference-in-difference estimation*. https://www.publichealth.columbia.edu/research/population-health-methods/difference-difference-estimation

Embassy of Japan in Peru. (2024, February 14). *RUNAi*. https://www.pe.emb-japan.go.jp/itpr_ja/11_000001_00720.html

Falaris, E. M. (2023). Children's school attendance, work, health and rainfall shocks in Ethiopia. *Applied Economics*, *55*(40), 4609–4624. https://www.tandfonline.com/doi/abs/10.1080/00036846.2022.2130147

Firmansyah, A. J., & Nurjani, E. (2023, August). The linkage of rainfall anomaly due to El Nino Southern Oscillation (ENSO) on cassava productivity in Central Java Province. *IOP Conference Series: Earth and Environmental Science*, *1233*(1), 012040. https://doi.org/10.1088/1755-1315/1233/1/012040

Flannery, M. J. (1986). Asymmetric information and risky debt maturity choice. *The Journal of Finance, 41*(1), 19–37.

FLAUC. (2017). *Fudan-Latin America University Consortium.* https://fddi.fudan.edu.cn/fddien/wudanwwatinwwmericawwniversity wwonsortiumwwwwwwww/list.htm

Haile, B. T., Zeleke, T. T., Beketie, K. T., Ayal, D. Y., & Feyisa, G. L. (2021). Analysis of El Niño Southern Oscillation and its impact on rainfall distribution and productivity of selected cereal crops in Kembata Alaba Tembaro zone. *Climate Services, 23,* 100254. https://ui.adsabs. harvard.edu/abs/2021CliSe.2300254H/abstract. https://doi.org/10.1016/ j.cliser.2021.10025

Index Mundi. (2024, February 13). *Sugar monthly price – US dollars per kilogram* [Commodity price]. https://www.indexmundi.com/commodities/? commodity=sugar&months=120

Instituto Del Mar Del Perú. (2024). *Official webpage of Instituto Del Mar del Perú.* https://www.gob.pe/imarpe

International Sugar Organization. (2017). *An evaluation of the global sugar market environment* [Report]. https://sugaralliance.org/wp-content/ uploads/2017/08/orive-2.pdf

Karuniasa, M., & Pambudi, P. A. (2022). The analysis of the El Niño phenomenon in the East Nusa Tenggara Province, Indonesia. *Journal of Water and Land Development, 52,* 180–185. https://doi.org/10.2478/ jwld-2022-0044

La República. (2017). *El Niño costero ha sido muy dañino para la industria azucarera Cámara de Comercio de la Libertad.* https://www. camaratru.org.pe/web2/index.php/2013/item/1800-el-nino-costero-ha-sid

Lal, D., & Singh, S. (2023). Impact of El-Nino and La-Nina episodes on rainfall variability and crop yield. *International Journal of Environment and Climate Change, 13*(10), 2046–2051. http://archiv.manuscptsubs. com/id/eprint/1650/

Li, Y., Strapasson, A., & Rojas, O. (2020). Assessment of El Niño and La Niña impacts on China: Enhancing the early warning system on food and agriculture. *Weather and Climate Extremes, 27,* 100208. https://doi. org/10.1016/j.wace.2019.100208

NOAA. (2024). *El Niño Southern Oscillation (ENSO)*. https://psl.noaa.gov/enso/

Noticias Trujillo. (2017). *UNT: Student volunteers, graduates and teachers clean university affected by the Huaico*. https://noticiastrujillo.pe/unt-huaico-limpieza/

OGCRI. (2024). *Oficina General de Cooperacion y Relaciones Interinstitucionales de la Universidad Nacional Mayor de San Marcos*. https://cooperacion.unmsm.edu.pe/#

Ojo Público. (2020). *Only 6% of the schools affected by the impacts of El Niño in 2017 have been rebuilt*. https://ojo-publico.com/edicion-regional/nino-costero-solo-se-ha-reconstruido-6-los-colegios-afectados

Pécastaing, N., & Chávez, C. (2020). The impact of El Niño phenomenon on dry forest-dependent communities' welfare in the northern coast of Peru. *Ecological Economics*, *178*, 106820. https://doi.org/10.1016/j.ecolecon.2020.106820

Pierce, H. F. (2017, March 23). *NASA examines Peru's deadly rainfall*. https://phys.org/news/2017-03-nasa-peru-deadly-rainfall.html

Ranganathan, A., & Benson, A. (2020). A numbers game: Quantification of work, auto-gamification, and worker productivity. *American Sociological Review*, *85*(4), 573–609.

Rotas, E., & Cahapay, M. (2020). Difficulties in remote learning: Voices of Philippine university students in the wake of COVID-19 crisis. *Asian Journal of Distance Education*, *15*(2), 147–158. https://www.asianjde.com/ojs/index.php/AsianJDE/article/view/504

Roth, J. (2022). Pretest with caution: Event-study estimates after testing for parallel trends. *American Economic Review: Insights*, *4*(3), 305–322. https://doi.org/10.1257/aeri.20210119

Sanabria, J., Bourrel, L., Dewitte, B., Frappart, F., Rau, P., Solis, O., & Labat, D. (2018). Rainfall along the coast of Peru during strong El Niño events. *International Journal of Climatology*, *38*(4), 1737–1747. https://doi.org/10.1002/joc.5288

Shah, M., & Steinberg, B. M. (2017). Drought of opportunities: Contemporaneous and long-term impacts of rainfall shocks on human capital. *Journal of Political Economy, 125*(2), 527–561. https://doi.org/10.1086/689162

Singh, M., Sah, S., & Singh, R. N. (2023). The 2023-24 El Niño event and its possible global consequences on food security with emphasis on India. *Food Security, 15*(6), 1431–1436. https://doi.org/10.1007/s12571-023-01419-8

SUNAT. (2024, February 15). *National superintendency of customs and tax administration.* https://www.sunat.gob.pe/

Thomalla, F., & Boyland, M. (2017). *Lessons from the 2015-2016 El Niño Event in Asia and the Pacific.* United Nations Development Program. https://www.unescap.org/sites/default/files/El%20Nino%20report-%20finalized%20ESCAP07082017.pdf

Touchstone, E. (2021). *How to ensure online education is a clear win for the SDGs Times Higher Education.* https://www.timeshighereducation.com/campus/how-ensure-online-education-clear-win-sdgs?cmp=1

United Nation. (2022). *Partnership in progress, Universidad Nacional Mayor de San Marcos and Liverpool University.* https://sdgs.un.org/partnerships/partnership-progress

United Nation. (2024, February 13). *End Hunger, achieve food security and improved nutrition and promote sustainable agriculture.* https://sdgs.un.org/goals/goal2#targets_and_indicators

United Nations. (2015). *United Nations millennium development goals.* https://www.un.org/millenniumgoals/poverty.shtml

United Nations. (2023). *Chapter 2 food security and nutrition around the world food and agricultural organization of the United Nations.* https://www.fao.org/3/cc3017en/online/state-food-security-and-nutrition-2023/food-security-nutrition-indicators.html

UNMSM. (2017). *Memory of Universidad Nacional Mayor de San Marcos 2017.* https://asambleauniversitaria.unmsm.edu.pe/archivos/MEMORIA-UNMSM_2017.pdf

UNMSM. (2024). *Official webpage of Universidad Nacional Mayor de San Marcos.* https://unmsm.edu.pe/

Vining, B. R., Hillman, A., & Contreras, D. A. (2022). El Niño Southern Oscillation and enhanced arid land vegetation productivity in NW South America. *Journal of Arid Environments, 198*, 104695.

World Bank. (2023, September 19). *Agriculture and food overview.* https://www.worldbank.org/en/topic/agriculture/overview

Yglesias-González, M., Valdés-Velásquez, A., Hartinger, S. M., Takahashi, K., Salvatierra, G., Velarde, R., Contreras, A., Santa María, H., Romanello, M., Paz-Soldán, V., Bazo, J., & Lescano, A. G. (2023). Reflections on the impact and response to the Peruvian 2017 Coastal El Niño event: Looking to the past to prepare for the future. *Plos One, 18*(9), e0290767. https://doi.org/10.1371/journal.pone.0290767

Zamudio, J. (2024a, April 9). *Internationalization experience.* https://www.jdzamudio.com/services/internacionalización

Zamudio, J. (2024b, February 16). *Teaching.* https://www.jdzamudio.com/services/teaching

4

FOOD SECURITY, NUTRITION, AND SUSTAINABLE AGRICULTURE NEXUS: THE ROLE OF HIGHER EDUCATION IN ATTAINMENT OF ZERO HUNGER IN ZIMBABWE

Prosper Chopera[a],
Tonderayi Mathew Matsungo[a],
Sandra Bhatasara[b], Viren Ranawana[c],
Alberto Fiore[d], Faith Manditsera[e] and
Lesley Macheka[b]

[a]University of Zimbabwe, Zimbabwe
[b]Marondera University of Agricultural Sciences and Technology, Zimbabwe
[c]University of Sheffield, UK
[d]Abertay University, UK
[e]Chinhoyi University of Technology, Zimbabwe

ABSTRACT

Sustainable development goal 2 (SDG2) is about creating a world free of hunger by 2030. Southern Africa faces a myriad of challenges affecting food and nutrition security, from population expansion, old and emerging pandemics, increased frequency of climate-induced natural disasters, ageing infrastructure, and challenging service delivery. The increased shocks and hazards and inadequate social safety nets have changed the dimensions of food and nutrition insecurity, giving rise to new roles for higher and tertiary education. Higher education (HE) institutions are expected to play a more active role in capacity building and producing goods and services that can contribute to the achievement of SDG2. This chapter assesses the role of HE towards the attainment of SDG2 which seeks to eliminate hunger and all forms of malnutrition. The chapter will highlight an insect-based value chain project as an example of HE contribution to reducing food insecurity. Through the case study, the chapter will explore the role of HE in community engagement, human capital development, and conducting research that informs policy and programming decisions. Furthermore, the chapter explores the role of North–South Collaborative research, interdisciplinary collaborations, and innovation hubs in developing innovations that can transform food systems and help build resilience in the face of the increasing climate and health shocks. Within these spaces, the contribution of HE to the achievement of food and nutrition security in Africa can be realised, and this approach replicated in other African institutions seeking to engage in such work.

Keywords: Hunger; food insecurity; insect protein; collaborations; higher education; innovation

INTRODUCTION

The United Nations (UN) adopted the sustainable development goals (SDGs) in September 2015 as the blueprint towards achieving 'a better and sustainable future for all' (United Nations, 2023a). Though 12 out of the 17 goals concern nutrition, the goal most directly committed to ending hunger, malnutrition, and achieving food security is SDG2 aptly named 'Zero hunger' (United Nations, 2023b).

Malnutrition, defined as deficiency or excess in nutrient intake affecting nutrition status (balance and utilisation of nutrients), has numerous aetiologies depending on the type concerned (Shetty, 2006). The UNICEF conceptual framework for malnutrition is held to be the best framework that elucidates the various pathways through which malnutrition becomes manifest (Black et al., 2020). The immediate causes of malnutrition are inadequate dietary intake and disease. Inadequate intake leads to physiological and biochemical changes at the cellular level impairing all or most biological systems (Emery, 2005). Then disease can cause excess loss of nutrients or impaired absorption and utilisation of nutrients in the body (Bhaskaram, 2002). These immediate causes have underlying causes that can be grouped into three domains: (i) food insecurity, (ii) inadequate care, and (iii) poor service delivery (nutrition, health, environment services, and social protection). Food insecurity can lead to inadequate dietary intake and consequently malnutrition (Avery, 2021). Inadequate child and maternal care will cause both inadequate dietary intake (Smith et al., 2005) and increased vulnerability to disease (Wagstaff et al., 2004), whilst poor services (nutrition, health, environmental, and social protection) can lead to higher disease vulnerability, frequent outbreaks, and high disease prevalence (Spencer, 2018). Beneath the underlying determinants are the basic causes of malnutrition that encompass resources (environmental, financial, etc.), politics, governance, and socio-cultural, religious and gender norms (Harris & Nisbett, 2021). The framework realises that to eradicate malnutrition all factors from basic causes to immediate causes must be addressed using a multi-sectoral approach as the causes of malnutrition are diverse and cross cutting (Keats et al., 2021). Though the framework was initially developed with an inclination towards undernutrition, developing countries are facing a double burden of malnutrition with overweight and obesity rising in prevalence (Wells et al., 2020). The causes of this other end of the spectrum have been attributed to nutrition transition (Popkin, 2001). This is the change from wholesome diets and active lifestyles to highly processed, high-calorie diets with sedentary lifestyles. Strong policies across the health and food system are required if overweight and obesity are to be combated (Hawkes et al., 2020; Keats et al., 2021).

Even with the knowledge of the causes of malnutrition, the goal of zero hunger eludes most stakeholders and policy makers. Progress in achieving SDG2 has been slow in most countries in Sub

Saharan Africa (SSA) for several reasons (Popkin, 2001). Africa is facing a lot of problems that threaten the achievement of SDG2 of zero hunger, for example, climate changes and environmental and socio-economic problems (Sokona & Denton, 2001). Climate change is a topical issue globally and is responsible for the environmental crisis being experienced the world over. While there is no agreed definition of the term, there is some consensus on the characteristics of climate change and its effects. Climate change poses a significant threat to Africa's food systems with the continent facing a range of climate change effects, such as increasing temperatures, changes in precipitation levels, and frequent droughts. Adverse weather conditions (droughts, soil erosion, etc.) over the years have meant reduced yields seriously affecting food security in SSA and a threat to the livelihoods of more than 1 billion people in SSA. This calls upon HE and policy makers to find innovative solutions to reduce the negative consequences on society and increase agricultural productivity-Target SDG2.3 and ensure sustainable food production systems with resilient agricultural practices-Target SDG2.4 (Adenle et al., 2017).

Adverse weather leading to rising sea levels, increased frequency of droughts, and famine have meant that farmers are producing less in the face of population expansion (Rust & Rust, 2013). Africa's population is estimated to be around 1.4 billion and is increasing 10-fold with a projection to reach 2.5 billion by 2050 (Guengant, 2017). This means that demographic pressures apart from climate change challenges will exacerbate food insecurity affecting the attainment of zero hunger. HE must investigate innovative solutions for improving food security, increasing water availability, and increasing the reliability of water available for agriculture and domestic purposes.

Water scarcity is not only unfavourable for agriculture but is undesirable at the household level too. For instance, water scarcity means compromised water and sanitation hygiene (WASH) and a rise in waterborne diseases. This affects the nutrition status of vulnerable households, such as wasting – low weight for height (Target SDG2.2) as continuously sick children fail to thrive adequately. However, if households are repeatedly exposed to recurring waterborne diseases, over time the children's health and dietary intake are negatively affected leading to high undernutrition and stunting rates (Targets SDG2.1 and 2.2).

Economic growth must be realised if SSA is to end hunger by 2030. Developing countries face a conundrum of socio-economic challenges all of which need a different approach and innovative solutions if these are to finally end. The current socio-economic challenges bedevilling developing countries include unstable economies, weak educational systems, poor health care systems, rising unemployment, and gender discrimination (Kyambalesa, 2004). Economic growth has a direct and indirect impact on attaining SDG2. Economic growth for instance can give rise to reduced unemployment rates and higher incomes that, in turn, may improve access to food. It can also mean governments have a broader tax base and can implement programmes with more social safety nets reducing the number of vulnerable households. Currently, HE institutions are pursuing innovation and industrialisation with the aim of encouraging economic growth and improvement of livelihoods (The World Bank, 2020). However, considering that budgets for higher and tertiary institutions are limited in Africa, there is heavy reliance on North–South collaborations to co-fund interventions such as The Insects4Nutrition Project described here.

CASE STUDY – INSECTS4NUTRITION (ZIMBABWE)

Edible insects are identified as the new protein sources with the potential to contribute to the food and nutrition security especially in developing countries (Selaledi et al., 2021). This is considering a growing world population and challenges brought about by climate change. Edible insects have been promoted for their valuable nutrient content, especially protein, and less greenhouse gas emissions for production (Van Huis & Dunkel, 2017). Due to their promising contribution towards nutrition security, there has been significant funding towards edible insects' research.

In 2020, the UK Biotechnology and Biological Sciences Research Council funded an international collaborative project on 'Upscaling edible insect-based porridge to improve health and nutritional status of primary school children in Zimbabwean low socio-economic communities' (UKRI, 2023). This is a collaborative effort involving the following universities: Abertay University (UK), Sheffield University (UK), the University of Zimbabwe, Marondera University of Agriculture Science and Technology (MUAST), and Chinhoyi University of Technology (CUT), a government department responsible for policy issues (The Food and Nutrition Council), food industry, and rural community cooperatives. The three-year project is aimed at optimising and upscaling the production process of a culturally acceptable innovative and novel insect-based food (porridge). Furthermore, the effectiveness of this developed mopane worm-based porridge on the nutritional outcomes of primary school children will be evaluated. This project addresses Target SDG2.1 of ending hunger and ensuring access to safe, nutritious food by vulnerable individuals all year round. Mopane worms are highly seasonal with limited success in studies attempting to break their diapause phase (Bara et al., 2022). In Zimbabwe, they only emerge during the February–May and October–December seasons. Attempts to domesticate them have failed due to the threat of parasites and diseases. However, value addition ensures they can be obtained throughout the year albeit in a different form (Potgieter et al., 2006). The project also addresses Target SDG2.2 of ending all forms of malnutrition as this porridge (a mix of mopane worm, pearl millet, and baobab) is high in protein, iron, and zinc and can improve the nutrition status of vulnerable children. Consumption of edible insects is a common practice among the rural and urban population in Zimbabwe (Manditsera et al., 2018).

In order to achieve its objectives, the project had the following work packages:

a. Evaluating the existing practices and the nutritional quality of the traditional and locally consumed edible insect-cereal porridge (Manditsera et al., 2022).
b. Developing a community acceptable and cheap nutrient dense mopane worm-based porridge with improved protein and mineral bioavailability for home use and industrial level (Ledbetter et al., 2024).

c. Evaluating the effects of daily supplementation of the insect-based porridge foods on linear growth and micronutrient status of primary school learners.

d. Upscaling the traditional mopane worms rearing technologies with communities being empowered to rear and process mopane worms.

Community Engagement

Indigenous knowledge systems are important in helping to identify the strategies that improve food and nutrition security and that can be upscaled. Firstly, the project evaluated the existing practices on edible insects and conducted an inventory of the available resources through community engagement. Community engagement involves first establishing relationships with all levels of local leadership and community members (from Provincial to District to Ward to Village and household level). The purpose and goals of the engagement are made clear, and mutual respect and trust are developed. Sometimes, establishing collaboration with the community that leads to capacity development and economic returns will strengthen relations. In the case of Insects4Nutrition, the raw material to produce the novel porridge was purchased from local cooperatives after training them on pre- and post-harvest procedures.

Nutritional analysis was conducted to determine the nutritional quality of the insects and other potential insects. Based on the information obtained from the community survey and the nutritional analysis, a mopane worm-based porridge was designed. The identified recipe was optimised in the laboratory in compliance with taste and child nutrient requirements (Ledbetter et al., 2024). The community was among the first to give feedback on the prototypes developed. It was vital to engage them in every step of the product development process and to give feedback on research results. Currently, the developed insect-based porridge is being fed to children to determine the impact of daily consumption of this high-protein, high-mineral porridge on the following child nutrition outcomes: iron status, wasting, underweight, and overweight

(Targets SDG2.1 and 2.2) in a randomised controlled trial. The last stage of the project will empower the community on various proven methods of value addition and will help establish a value chain stakeholder network platform that links community harvesters/ traders with markets.

This project demonstrates the role of HE institutions in working with communities to promote food and nutrition security, thus reducing hunger and the prevalence of malnutrition (Targets SDG2.1 and 2.2) through traditional food systems research and innovation. For this project, the university was an intermediary between industry and the communities where the raw materials (mopane worms and climate smart cereals) are continuously being sourced. The HE researchers' role was to document the existing practices, critique the status quo, and identify ways in which local resources can be fully exploited for improved nutrient bioavailability and enhanced sensorial properties in a value chain approach. However, due to infrastructural limitations, the university needed to collaborate with industry for upscaling of the developed prototype. The Insects4Nutrition project also demonstrates the importance of interdisciplinary collaboration with researchers consisting of food scientists, nutritionists, sociologists, and policy makers. These partnerships and collaborations are of importance to achieving SDG2. Researchers in HE and industry must work synergistically to improve the food and nutrition security of vulnerable groups. Key and central to this collaboration are the existence of innovation hubs as well as North–South collaborations.

HE INNOVATION HUBS AND THE ATTAINMENT OF ZERO HUNGER

Innovation hubs are an 'incubation space which seek to ignite social change and unlock the potential of innovators in contributing to sustainable development through novel ideas' (Grants and Resources for Sustainability, 2016). Innovation hubs help to facilitate the transformation of promising ideas into practical solutions that address real social challenges faced by local communities in which the young people are resident (Berger & Brem, 2016).

Over the last two decades, there has been an increase in the development of innovation hubs within the education sector worldwide. Countries have invested in their creation with the aim of fostering a 'technopreneurial' culture based on innovation and research, and directly stimulating and supporting innovative start-ups (Joshua & More, 2022). Innovation hubs have been emerging across Africa since the first one opened in South Africa in 2005 (Baark & Sharif, 2006). Zimbabwe has now joined countries like Kenya, Nigeria, Eswatini, Rwanda, and South Africa making use of innovation hubs in driving innovation and industrialisation.

Food and agricultural innovations play essential roles in overcoming hunger sustainably and effectively (Tomich et al., 2019; Zwane, 2020). Hunger is a complex problem, and evidence from HE and innovation hubs is helping to unravel this conundrum and help find effective and cost-efficient solutions to the problem. However, innovation is not just a matter of technology and productivity. Scaling up of innovations requires capital and platforms to bring together relevant stakeholders to enable the adoption process (transfer of technology and know-how). It is here that specialised food system innovation hubs have a key role to play. Food system innovation hubs can provide transformative solutions to food systems by bringing the right innovations to market faster in a cost-effective manner (Reardon et al., 2019). The lessons learnt from the Insects4Nutrition project in working with and under food system innovation hubs include the following requirements, in that there needs:

- a clear reward system in place to motivate innovators;

- a clear policy on intellectual property protection;

- adequate and consistent funding or most innovations will not develop further from prototype stage;

- the roles, responsibilities of each stakeholder must be clear. Stakeholders here include the department, the faculty, the innovator, the hub, the university. Overlapping of roles and overstepping mandates can lead to duplication of roles, frustration, and confusion.

Food system innovation hubs rely on collaboration with key agencies, public–private partnerships, and social relations with the global North playing an important role in evidence generation.

NORTH–SOUTH COLLABORATIONS IN HE RESEARCH AND SDG2

Never has the need for collaboration between the global North and South been more important than now as the world navigates unprecedented global challenges dominated by climate change, food security, and health. Although, historically, food security nutrition and sustainable agriculture were typically seen as problems of the global South, these are now rapidly emerging as major issues also for the global North (Savary et al., 2022). This emphasises the need to rethink conventional perspectives on HE to one of more equitable learning across the North–South spectrum. Addressing gaps in HE capabilities and skills is essential for facilitating equitable learning and partnerships.

Historically, global HE has been one-sided with the global South seen as recipients and the North as providers, driven by differentials in power relations and in socio-economic development (Walker & Martinez-Vargas, 2022). The latter remains true for many countries in the global South, and aid is still an important factor for achieving socio-economic equity, including improving HE. Recent Organisation for Economic Cooperation and Development (OECD) data show official donors provide around US$200 billion as aid for developing countries (OECD, 2023), with Africa receiving the largest proportion of these funds at around US$76 billion. International development of HE emphasises the benefits of investing in the sector to improve human capital, economies, civil societies, and incomes (Chankseliani, 2022). Therefore, the continued use and allocation of foreign aid for funding HE initiatives to reduce inequalities should be a priority in North–South collaborations.

The UK funds research in the global south through its public funder of research, UK Research and Innovation (UKRI), using funds allocated for official development assistance. Schemes such as the Global Challenges Research Fund and the Newton Fund are dedicated to funding North–South global research and have been pivotal for driving food security, nutrition, and sustainable agriculture research between the North and South.

The Insects4Nutrition project a recipient of UKRI funding has been able to achieve a lot of its objectives to date. The funding has enabled collaboration and knowledge exchange between the collaborating universities from the global north (University of Abertay, Scotland, and Sheffield University, UK) and the University of Zimbabwe, Chinhoyi University of Technology, and Marondera University of Agriculture, Science and Technology all in Zimbabwe. The technical and financial support has facilitated the creation of stakeholder networks between academia, food industry in this case National Foods who have upscaled production of the extruded product, policy (Food and Nutrition Council), and the communities in the South of Zimbabwe. This has been essential in driving the agenda on food and nutrition security alleviation.

While degrees and scholarly learning are important elements of HE, there is an equal and, in some cases, more pressing need for training on vocational and technical skills essential for addressing food security, nutrition, and sustainable agriculture challenges. Skills training is included as a mandatory component in some grant calls, and this is a positive move by funders. However, there is an urgent need for more funding dedicated to vocational and technical skills training at HE level.

A noteworthy barrier to levelling differences between the global North and South and enabling opportunities for HE is the digital divide (Mammen et al., 2023). Data show that only 36% of Africa's population has broadband access with even less access in poorer and remote regions (World Bank, 2023). Africa also shows the greatest gender disparities in internet use with only around 24% of women using it. Improving digital literacy and digital access is key for facilitating HE and global collaboration and should be a priority for governments and policy. North–South collaborations are needed for addressing barriers to digital access such as poor infrastructure, affordability, and digital literacy skills. Despite the inequalities, a paradigm shift in HE collaborations is needed that steers away from North–South divides to one that fosters a global culture of equal partnerships. This would be pivotal for promoting two-way learning and growing a global human capital that can work as one at the food security, nutrition, and sustainable agriculture nexus.

GENDER AND FOOD-SYSTEMS INNOVATIONS

Innovations should be inclusive and produce equitable outcomes for various social groups. It is, therefore, essential to promote Gendered Food Systems Innovations. These are processes that integrate gender analysis into all phases of the innovation cycle to ensure that the outcomes of the innovation benefit all or do not create unintended negative consequences on a specific group based on their gender (Me-Nsope, 2017). Gender is a 'social construct that refers to the array of socially constructed roles and relationships, personality traits, attitudes, behaviours, values, and relative power and influence that society ascribes to the two sexes on a differential basis' (Me-Nsope, 2017). Gender can be both an accelerator and inhibitor of innovations by HE institutions. Hence, if not well articulated in especially policy frameworks it will not be addressed well to the detriment of food and nutrition security of vulnerable groups. The positive relationship between food security, nutrition, and gender equality means it is imperative to systematically consider gender power relations in the contexts in which they operate. Gender analysis is simply defined as 'a process to identify, understand, and explain gaps between males and females that exist in households, communities, and countries' (European Commission, 2002). Understanding who plays what role in providing food security at the household level is crucial for targeting interventions that aim to increase agricultural productivity (Target SDG2.3), end hunger (Target SDG2.1), and eradicate malnutrition (Target SDG2.2) in communities.

Furthermore, developing innovations to close gender gaps and tackle the structural causes of inequalities in food and nutrition security requires high-quality research and data disaggregated by sex, age, and other dimensions of social and economic differentiation (FAO, 2023). As such, research by HE should be grounded in systematic gender analysis. The World Bank's Gender Innovation Lab has found that working with all genders can yield enhanced outcomes in several circumstances including increased productivity in specific value chains (World Bank & World Trade Organisation, 2020). It has also been noted that co-designing and bundling innovations could offer significant opportunities to achieve better outcomes (Johnstone et al., 2023). Others have also noted that to empower women in a sustainable manner in the food system, education and training for girls and women on STEM subjects (science, technology, engineering, and math) and other relevant subjects such as nutrition education, entrepreneurship, and on modern farming techniques and technologies is pertinent (Malabo Montpellier Panel, 2023). HE institutions, thus, provide a fertile ground for this training and capacity building. There is evidence showing positive effects of nutrition counselling, nutrition education, and maternal education for nutrition, dietary diversity, and health outcomes for women and children (Choudhury et al., 2020). Specifically in Zimbabwe, the 'Caregroup' approach, a women's support group encouraging peer-to-peer learning, is being promoted in many disadvantaged rural communities and has led to many positive outcomes in women and child nutrition indicators (Macheka et al., 2022; Matsungo et al., 2023).

HE innovation hubs can serve to promote gender equality and women's empowerment by improving availability and access to safe and nutritious food. Women's access to safe and nutritious food is impacted by various factors including income, with women earning less. HE can intervene by providing novel low cost safe and nutritious food, as the case study on Insects4Nutrition demonstrates. Another key issue is for HE innovation hubs to create networks and linkages to markets where different socio-economic groups can purchase nutritious food. Faculties of agriculture and their departments can spearhead research, support, and promote gender-responsive agricultural innovations. Overall innovations are found to be more effective when they are cross cutting and include components on nutrition and health behaviour change communication (SDG3 – *good health and wellbeing*); women's empowerment (SDG5 – *gender equality* and SDG10 – *reduced inequalities*); water, sanitation, and hygiene (WASH) (SDG6 – *clean water and sanitation*); and micronutrient-fortified products (SDG2) (Ruel et al., 2018).

CONCLUSIONS AND RECOMMENDATIONS

This chapter illustrates an example of the role of HE in the attainment of SDG2: *end hunger, achieve food security and improved nutrition, and promote sustainable agriculture* particularly Targets SDG2.1 to 2.4. SDG2 links food security, nutrition, and a sustainable but climate resilient agriculture. Our narrative was on the African context with glimpses of the need for North–South collaborations to ensure the maximum impact of HE initiatives and/or research. A case study was presented highlighting how an ongoing collaborative research project between Abertay University (UK), Sheffield University (UK), University of Zimbabwe, MUAST, and CUT is contributing to improving the food and nutrition security in Zimbabwe.

SDG2A specifically mentions the need to increase investment, including through enhanced international cooperation, in rural infrastructure, agricultural research and extension services, technology development, and plant and livestock gene banks in order to enhance agricultural productive capacity in developing countries, in particular least developed countries (United Nations, 2023b). Thus, SDG2A, in particular, is the target that HE will contribute more towards its attainment.

Overall, there is a need for research that understands both social issues and science to provide evidence-based novel solutions to prevailing food and nutrition security problems in Africa (Ashida, 2023). It is critical to disseminate scientific findings to society in an easy-to-understand manner and to promote dialogue between science and society, such that an emphasis on scientific communication and capacity-building training should equip graduates with the necessary skills to address the community challenges. In this regard, HE institutions are called to lead the global agenda for agriculture and food system transformation. Compared to the past, a more proactive approach is now being taken to achieve the SDGs with industry partners being considered 'a vital cog in the machine' especially for scaling up of interventions (Ashida, 2023). Transforming food systems means raising standards, ensuring that food systems provide healthy diets and living incomes to everyone involved in the food value chain in some way and that they do this in an environmentally sustainable way (Dengerink et al., 2022).

HE has a critical role to play in this food security, agriculture, and nutrition nexus, and our key recommendations highlight

1. The need for training of graduates that drive R&D innovations.

2. The promotion of scientific research to provide evidence-based solutions to prevailing country-specific food and nutrition security problems.

3. To achieve recommendations 1 & 2 calls for deliberate efforts to strengthen the technical capacity of HE and research institutions so that they themselves can transfer knowledge to communities towards poverty eradication and sustainable development.

4. The importance of funding for research and innovation and partnerships with industry and collaborations with global North partners and/or development partners to drive relevant innovations is critical.

The lessons learned from implementing the multi-partner Insects4Nutrition project in Zimbabwe, such as appreciation for indigenous knowledge systems, community engagement, technical and financial support, gender equality, and key collaborations, are valuable insights for scaling up efforts to tackle SDG2. In summary, all HE and research institutions should leverage on strategic partnerships and creatively transfer their knowledge, skills, and innovations to evidence-based practice to drive sustainable community development within the spirit of the SDGs of 'Leaving no one and no place behind' (UN Sustainable Development Group, 2024).

REFERENCES

Adenle, A. A., Ford, J. D., Morton, J., Twomlow, S., Alverson, K., Cattaneo, A., Cervigni, R., Kurukulasuriya, P., Huq, S., Helfgott, A., & Ebinger, J. O. (2017). Managing climate change risks in Africa – A global perspective. *Ecological Economics, 141,* 190–201. https://doi.org/10.1016/j.ecolecon.2017.06.004

Ashida, A. (2023). The role of higher education in achieving the sustainable development goals. In S. Urata, K. Kuroda, & Y. Tonegawa (Eds.), *Sustainable development disciplines for humanity: Breaking down the 5Ps – People, planet, prosperity, peace, and partnerships* (pp. 71–84). Springer Nature Singapore. https://doi.org/10.1007/978-981-19-4859-6_5

Avery, A. (2021). Food insecurity and malnutrition. *Kompass Nutrition & Dietetics, 1*(2), 41–43. https://doi.org/10.1159/000515968

Baark, E., & Sharif, N. (2006). From trade hub to innovation hub: The role of Hong Kong's innovation system in linking China to global markets. *Innovation, 8*(1–2), 193–209. https://doi.org/10.5172/impp.2006.8.1-2.193

Bara, G., Sithole, R., & Macheka, L. (2022). The mopane worm (Gonimbrasia belina Westwood): A review of its biology, ecology and utilisation in Zimbabwe. *Journal of Insects as Food and Feed, 8*(8), 823–836. https://doi.org/10.3920/jiff2021.0177

Berger, A., & Brem, A. (2016). Innovation hub how-to: Lessons from silicon valley. *Global Business and Organizational Excellence, 35*(5), 58–70. https://doi.org/10.1002/joe.21698

Bhaskaram, P. (2002). Micronutrient malnutrition, infection, and immunity: An overview. *Nutrition Reviews, 60*(suppl_5), S40–S45. https://doi.org/10.1301/00296640260130722

Black, M. M., Lutter, C. K., & Trude, A. C. B. (2020). All children surviving and thriving: Re-envisioning UNICEF's conceptual framework of malnutrition. *The Lancet Global Health, 8*(6), e766–e767. https://doi.org/10.1016/S2214-109X(20)30122-4

Chankseliani, M. (2022). International development higher education: Looking from the past, looking to the future. *Oxford Review of Education, 48*(4), 457–473. https://doi.org/10.1080/03054985.2022.2077325

Choudhury, S., Shankar, B., Aleksandrowicz, L., Tak, M., Green, R., Harris, F., Scheelbeek, P., & Dangour, A. (2020). What underlies inadequate and unequal fruit and vegetable consumption in India? An exploratory analysis. *Global Food Security, 24*, 100332. https://doi.org/10.1016/j.gfs.2019.100332

Dengerink, J., Piters, B., de S., Brouwer, H., & Guijt, J. (2022). *Food systems transformation: An introduction.* Wageningen University & Research.

Emery, P. W. (2005). Metabolic changes in malnutrition. *Eye, 19*(10), 1029–1034. https://doi.org/10.1038/sj.eye.6701959

European Commission. (2002). *Report on the communication from the commission to the council and the European Parliament on the programme of action for the mainstreaming of gender equality in community development cooperation (Hughes procedure) – Committee on women's rights and equal opportunities | A5-0066/2002 | European Parliament.* https://www.europarl.europa.eu/doceo/document/A-5-2002-0066_EN.html?redirect

FAO. (2023). *The status of women in agrifood systems.* https://doi.org/10.4060/cc5343en

Grants and Resources for Sustainability. (2016). *FundsforNGOs.* https://www.fundsforngos.org/developing-countries-2/zimbabwe-2/green-innovation-hub-gihub-submit-social-innovating-ideas-for-environmental-sustainability-and-renewable-energy-needs/

Guengant, J.-P. (2017). Africa's population: History, current status, and projections. In H. Groth & J. F. May (Eds.), *Africa's population: In search of a demographic dividend* (pp. 11–31). Springer International Publishing.

Harris, J., & Nisbett, N. (2021). The basic determinants of malnutrition: Resources, structures, ideas and power. *International Journal of Health Policy and Management, 10*(Special Issue on Political Economy of Food Systems), 817–827. https://doi.org/10.34172/ijhpm.2020.259

Hawkes, C., Ruel, M. T., Salm, L., Sinclair, B., & Branca, F. (2020). Double-duty actions: Seizing programme and policy opportunities to address malnutrition in all its forms. *The Lancet, 395*(10218), 142–155.

Johnstone, K., Thazin Aung, M., & Barrett, S. (2023). *Can innovations in agri-food systems deliver gender equity and resilience?* https://www.iied. org/21591iied

Joshua, S., & More, C. (2022). A stakeholder founded business model for strategic management of innovation hubs: A case of Zimbabwe universities innovation hubs. *Journal of African Education, 3*(2), 155–179. https://doi.org/10.31920/2633-2930/2022/v3n2a6

Keats, E. C., Das, J. K., Salam, R. A., Lassi, Z. S., Imdad, A., Black, R. E., & Bhutta, Z. A. (2021). Effective interventions to address maternal and child malnutrition: An update of the evidence. *The Lancet Child & Adolescent Health, 5*(5), 367–384.

Kyambalesa, H. (2004). *Socio-economic challenges: The African context.* Africa World Press.

Ledbetter, M., Wilkin, J. D., Mubaiwa, J., Angeline Manditsera, F., Macheka, L., Matiza Ruzengwe, F., Madimutsa, O. N., Chopera, P., Matsungo, T. M., Cottin, S. C., Stephens, E. W., Ranawana, V., & Fiore, A. (2024). Development of a nutritious cereal-based instant porridge by the incorporation of protein-rich insect powder – An example from Zimbabwe. *Journal of Functional Foods, 112,* 105957. https://doi. org/10.1016/j.jff.2023.105957

Macheka, L., Bhatasara, S., Mugariri, F., Takawira, D., Kairiza, T., & Matsungo, T. M. (2022). Impact of care group participation on nutrition knowledge, behaviour and practices. *The North African Journal of Food and Nutrition Research, 6*(13), 87–93. https://doi.org/10.51745/ NAJFNR.6.13.87-93

Malabo Montpellier Panel. (2023). *Bridging the gap: Policy innovations to put women at the centre of food systems transformation in Africa.* https://www.mamopanel.org/resources/women-agrifood-systems/

Mammen, J. T., Rugmini Devi, M., & Girish Kumar, R. (2023). North–South digital divide: A comparative study of personal and positional inequalities in USA and India. *African Journal of Science, Technology, Innovation and Development, 15*(4), 482–495. https://doi.org/10.1080/ 20421338.2022.2129343

Manditsera, F. A., Lakemond, C. M. M., Fogliano, V., Zvidzai, C. J., & Luning, P. A. (2018). Consumption patterns of edible insects in rural and urban areas of Zimbabwe: Taste, nutritional value and availability are key elements for keeping the insect eating habit. *Food Security, 10*(3), 561–570. https://doi.org/10.1007/s12571-018-0801-8

Manditsera, F. A., Mubaiwa, J., Matsungo, T. M., Chopera, P., Bhatasara, S., Kembo, G., Mahlatini, H., Matiza Ruzengwe, F., Matutu, F., Grigor, J., Fiore, A., & Macheka, L. (2022). Mopane worm value chain in Zimbabwe: Evidence on knowledge, practices, and processes in Gwanda District. *PLOS One, 17*(12), e0278230. https://doi.org/10.1371/journal.pone.0278230

Matsungo, T. M., Kamazizwa, F., Mavhudzi, T., Makota, S., Kamunda, B., Matsinde, C., Chagwena, D., Mukudoka, K., & Chopera, P. (2023). Influence of care group participation on infant and young child feeding, dietary diversity, WASH behaviours and nutrition outcomes in rural Zimbabwe. *BMJ Nutrition, Prevention & Health, 6*(2), 164. https://doi.org/10.1136/bmjnph-2023-000627

Me-Nsope, N. (2017). *Gender practice in food system innovations: Approaches lessons and challenges.* gcfsi.isp.msu.edu/files/8815/1207/2281/Gender_Practice_Food_Systems_Innovations_Nov._2017.pdf

OECD. (2023). *AID (ODA) disbursements to countries and regions.* https://stats.oecd.org/Index.aspx?DataSetCode=Table2A

Popkin, B. M. (2001). The nutrition transition and obesity in the developing world. *The Journal of Nutrition, 131*(3), 871S–873S.

Potgieter, M., Mushongohande, M., & Wessels, D. (2006). Mopane tree ecology and management. In J. Ghazoul (Ed.), *Mopane woodlands and the Mopane worm: Enhancing rural livelihoods and resource sustainability technical report* (pp. 7–17). DFID.

Reardon, T., Echeverria, R., Berdegué, J., Minten, B., Liverpool-Tasie, S., Tschirley, D., & Zilberman, D. (2019). Rapid transformation of food systems in developing regions: Highlighting the role of agricultural research & innovations. *Agricultural Systems, 172*, 47–59. https://doi.org/10.1016/j.agsy.2018.01.022

Ruel, M. T., Quisumbing, A. R., & Balagamwala, M. (2018). Nutrition-sensitive agriculture: What have we learned so far? *Global Food Security*, *17*, 128–153. https://doi.org/10.1016/j.gfs.2018.01.002

Rust, J. M., & Rust, T. (2013). Climate change and livestock production: A review with emphasis on Africa. *South African Journal of Animal Science*, *43*(3), 255–267.

Savary, S., Waddington, S., Akter, S., Almekinders, C. J. M., Harris, J., Korsten, L., Rötter, R. P., & Van den Broeck, G. (2022). Revisiting food security in 2021: An overview of the past year. *Food Security*, *14*(1), 1–7. https://doi.org/10.1007/s12571-022-01266-z

Selaledi, L., Hassan, Z., Manyelo, T. G., & Mabelebele, M. (2021). Insects' production, consumption, policy, and sustainability: What have we learned from the indigenous knowledge systems? *Insects*, *12*(5), 432. https://doi.org/10.3390/insects12050432

Shetty, P. (2006). Malnutrition and undernutrition. *Medicine*, *34*(12), 524–529. https://doi.org/10.1053/j.mpmed.2006.09.014

Smith, L. C., Ruel, M. T., & Ndiaye, A. (2005). Why is child malnutrition lower in urban than in rural areas? Evidence from 36 developing countries. *World Development*, *33*(8), 1285–1305. https://doi.org/10.1016/j.worlddev.2005.03.002

Sokona, Y., & Denton, F. (2001). Climate change impacts: Can Africa cope with the challenges? *Climate Policy*, *1*(1), 117–123. https://doi.org/10.3763/cpol.2001.0110

Spencer, N. (2018). The social determinants of child health. *Paediatrics and Child Health*, *28*(3), 138–143. https://doi.org/10.1016/j.paed.2018.01.001

The World Bank. (2020). *Revitalising Zimbabwe's tertiary education sector to support a robust economic recovery*.

Tomich, T. P., Lidder, P., Coley, M., Gollin, D., Meinzen-Dick, R., Webb, P., & Carberry, P. (2019). Food and agricultural innovation pathways for prosperity. *Agricultural Systems*, *172*, 1–15. https://doi.org/10.1016/j.agsy.2018.01.002

UKRI. (2023). *Upscaling edible insect based porridge to improve health and nutrition status of Primary school children in Zimbabwean low*

socio economic communities. https://gtr.ukri.org/projects?ref=BB%2FT0
09055%2F1

United Nations. (2023a). *The 17 goals. Sustainable development.
Sustainable development goals*. https://sdgs.un.org/goals

United Nations. (2023b, November 18). *Goal 2 Zero Hunger. Sustainable
development goals*. https://un.org/sustainabledevelopment/hunger/

UN Sustainable Development Group. (2024). *Leave no one behind*.
United Nations Sustainable Development Group. https://unsdg.
un.org/2030-agenda/universal-values/leave-no-one-behind

Van Huis, A., & Dunkel, F. V. (2017). Edible insects: A neglected and
promising food source. *Sustainable Protein Sources, 2017*, 341–355.
https://doi.org/10.1016/B978-0-12-802778-3.00021-4

Wagstaff, A., Bustreo, F., Bryce, J., & Claeson, M. (2004). Child health:
Reaching the poor. *American Journal of Public Health*, 94(5), 726–736.
https://doi.org/10.2105/AJPH.94.5.726

Walker, M., & Martinez-Vargas, C. (2022). Epistemic governance and the
colonial epistemic structure: Towards epistemic humility and transformed
South-North relations. *Critical Studies in Education*, 63(5), 556–571.
https://doi.org/10.1080/17508487.2020.1778052

Wells, J. C., Sawaya, A. L., Wibaek, R., Mwangome, M., Poullas, M. S.,
Yajnik, C. S., & Demaio, A. (2020). The double burden of malnutrition:
Aetiological pathways and consequences for health. *The Lancet*,
395(10217), 75–88.

World Bank. (2023). *From connectivity to services: Digital transformation
in Africa*. https://www.worldbank.org/en/results/2023/06/26/
from-connectivity-to-services-digital-transformation-in-africa/

World Bank & World Trade Organisation. (2020). *Women and trade: The
role of trade in promoting gender equality*. https://www.wto.org/english/
res_e/booksp_e/women_trade_pub2807_e.pdf

Zwane, E. (2020). The role of agricultural innovation system in
sustainable food security. *South African Journal of Agricultural Extension*,
48(1), 122–134. https://doi.org/10.17159/2413-3221/2020/v48n1a531

5

CONTRIBUTION OF LILONGWE UNIVERSITY OF AGRICULTURE AND NATURAL RESOURCES TO THE ATTAINMENT OF SDG2 IN MALAWI

Agnes Mbachi Mwangwela, Vincent Mlotha,
Alexander Archippus Kalimbira,
William Kasapila, Jessica Kampanje Phiri,
Samuel Mwango, Samson Pilanazo Katengeza

Lilongwe University of Agriculture and Natural Resources, Malawi

ABSTRACT

A case study of Lilongwe University of Agriculture and Natural Resources (LUANAR) in Malawi explores its contribution to improving food security and nutrition using varied genetic resources and plant-based diets. The chapter articulates specific examples of research and outreach activities conducted to improve availability, access, and consumption of safe and quality food to reduce undernutrition. Malawi, together with other countries, adopted the global 2030 sustainable development goals (SDGs) during the United Nations General Assembly in September 2015 to transform the world, end poverty and inequality, protect the planet, and ensure that all people enjoy health, justice, and prosperity. SDG2 is on ending hunger, achieving food security, improving nutrition, and promoting sustainable agriculture. Malawi has made significant progress and is on track to achieving SDG number 2 by 2030, and LUANAR has contributed to this achievement in multiple ways. The university has academic programmes and carries out research in various areas of agriculture and natural resources that relate directly to SGD 2. The faculty of Food and Human Sciences champions training, research, and innovation on food and nutrition at the university. The chapter concludes by reiterating that government leadership, support from development partners, and collaboration with the academic, research, and private sectors are key to success. The research models, impact, and challenges presented in the chapter have relevance and potential for wider application in the developing world.

Keywords: Sustainable food systems; university and higher education; sustainable development goals; developing countries; nutrition and health; Malawi

1. INTRODUCTION

In September 2015, Malawi was one of the 186 United Nations (UN) Member States that adopted and ratified the 2030 agenda for sustainable development. The agenda contains 17 sustainable development goals – SDGs, which focus on fostering peace and prosperity for citizens in all member states of the UN and enhancing global partnership for development for developed and developing member states. The 2030 agenda hinges on boosting economic growth through ending poverty and various forms of deprivations, tackling climate change, and preserving natural resources. Principles behind the agenda include people, planet, prosperity, peace, and partnership.

The SDGs require key transformation of key sectors of development and leaving no one behind, starting with the furthest behind. To achieve the SDGs, a high level of cooperation from member states and local mobilisation of resources are necessary. The Government of

Malawi (GoM) is domesticating the SDGs in its 10-year (2021–2022) national development plan (MIP-1) of the Malawi Vision 2063 (MW2063), which has replaced the Malawi Growth and Development Strategy III of 2017–2022. Launched in January 2021, the MW2063 charts Malawi's new development trajectory of achieving an upper middle-income status by the year 2063 (Government of Malawi (GoM), 2021). The MW2063 reflects collective aspirations of the people of Malawi to achieve inclusive wealth creation and self-reliance for economic independence. The MIP-1 is anchored on the three pillars of MW2063, namely, agricultural productivity and commercialisation; industrialisation; and urbanisation. It has an average alignment of 81.62% to the SDGs (Table 5.1).

1.1. Achievement of the SDGs in Malawi

The GoM has so far produced one progress report published in 2018 and two voluntary national review (VNR) reports in 2020 and 2022 (GoM, 2022). The VNR process involves countries assessing themselves on SDG implementation and sharing findings with fellow member states of the UN for peer review. The 2022 VNR report shows that Malawi has registered significant progress on SDG2: zero hunger; The African SDGs index and dashboard ranks Malawi 25 out of 52 countries on the continent, while globally Malawi is in position 146 out of 162 countries (Sachs et al., 2016).

1.1.1. Sustainable Development Goal (SDG) 2: Zero Hunger

Malawi is on track to achieve zero hunger by 2030. However, frequent episodes of hunger and food insecurity experienced every year are a warning that there is considerable work to be done to make sure the country 'leaves no one behind' (World Food Program (WFP), 2020) on the road towards a world with zero hunger. It is unlikely there will be zero hunger by 2030 (going by the current trends) if the adverse effects of climate change continue. However, the country is making good progress in its efforts to end the problem, evidenced by a score of 21.3% on the Global Health Index

Table 5.1. Synergy Between MW2063 and Agenda 2030 for SDGs.

MW2063 Pillar	Priority Strategies, Interventions, and Game Changers	SDGs to be Achieved
Pillar 1 (agricultural productivity and commercialisation): to have an optimally productive and commercialised agriculture sector	Agriculture diversification, irrigation development, anchor farms, agriculture inputs, agricultural mechanisation, farm inputs and agriculture research, innovation and dissemination	Goals 1 (no poverty), 2 (zero hunger), 3 (good health and well-being), and 13 (climate action)
Pillar 2 (industrialisation): to have a vibrant knowledge-based economy with a strong manufacturing industry driven by productive and commercially vibrant agriculture and mining sectors	Industrialising mining: diversification, value addition, and competitiveness largely targeting the export market; creation of special economic and export processing zones; and research, science, technology, and innovation	Goals 8 (decent work and economic growth), 12 (responsible consumption and production)
Pillar 3 (urbanisation): To have world-class urban centres and tourism hubs across the country, with the requisite modern socio-economic amenities	Development of secondary cities, infrastructure development planning and investment in cities and towns that are regulated and controlled, sustainable municipal self-financing mechanisms and tourism development	SDG4 (quality education), 6 (clean water and sanitation), 8 (decent work and economic growth), 9 (industry, innovation, and infrastructure), 11 (sustainable cities and communities)

Severity Scale compared to the regional average of 27.1% (GoM, 2022). These improvements are due to improved food security owing to increased agricultural productivity in recent years. For instance, in 2020/2021, almost all crops recorded significant increments in production with maize registering about 21% increase. Despite such strides, the VNR of 2022 reveals that poor people continue to face some challenges in accessing quality food at the community level. The effects of climate change and the general rise in commodity prices are key risks to the achievement of this goal.

1.1.2. Indicator 2.1.1 Prevalence of Undernourishment

Between 2000 and 2019, Malawi reduced prevalence of undernourishment by 19.8% from 20.7 to 17.3%. Significantly, the number of people vulnerable to perennial hunger has declined. As of the 2021/2022 farming seasons, 1,496,396 people (8% of the population) required food assistance, a 43% decrease from the preceding year.

1.1.3. Indicator 2.2.1 Prevalence of Stunting Among All Children Under Five Years of Age

Stunting, in children under five years of age, has reduced from 55% in 1992 and 2000 to 53% in 2004, 47% in 2010 and reducing by 10 percentage points for the first time to 37.1% in 2015/2016 (National Statistical Office (NSO) [Malawi] and ICF, 2017) after the country adopted the scaling up nutrition (SUN) policy brief. The recent Malawi Multiple Indicator Cluster Survey of 2019 further reported a 35.5% stunting prevalence, which has decreased to 33.7% (GoM, 2019). For children under five years of age, the World Health Organisation (WHO) sets a low prevalence of stunting at <20%, while 20–29% indicates medium, 30–39% high and >40% very high prevalence. The current rate in Malawi is, therefore, still high and needs to be reduced further. Fig. 5.1 shows the trends in prevalence of stunting among children under five years of age in Malawi from 1992 to 2020.

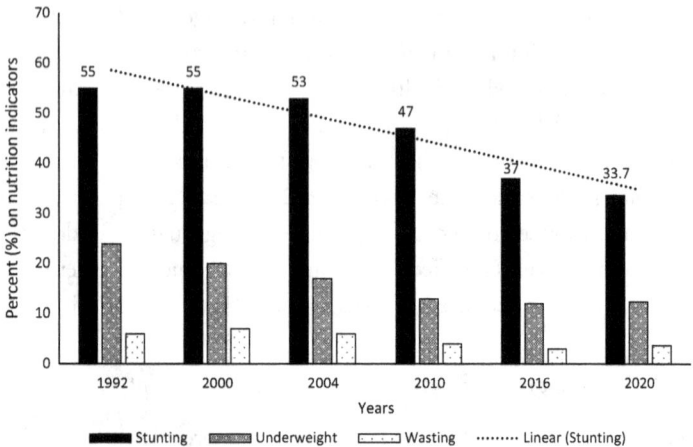

Fig. 5.1.   Separate File.

1.1.4. Indicator 2.2.2 Prevalence of Malnutrition Among Under-Five Children

The WHO thresholds for underweight are <10%, low; 10–19%, medium; 20–29%, high; and >30%, very high prevalence. For wasting, they are <5% acceptable, 5–9% poor, 10–14% serious, and 15% or more critical. The prevalence of overweight and wasting have been declining while underweight has worsened between 2016 and 2019 (GoM, 2019; National Statistical Office (NSO) [Malawi] and ICF, 2017). Despite a marginal increase in the prevalence of underweight, the long-run trend has been declining from 2000 (20%) to 2019 (12.8%). The prevalence took a dip from 11.7% recorded in 2016.

1.1.5. Interventions Under Goal 2

The increase in agricultural productivity and food security and reduction in undernutrition in the country has been because of

(1) The affordable input programme that subsidises farm inputs for poor households and has led to an increase in agricultural productivity.
(2) Intensified irrigation development increased the areas under irrigation by 22.5% since the last VNR.

(3) Enhanced implementation of nutrition programmes through stakeholders such as community-based management of acute malnutrition, treatment of severe acute malnutrition, infant and young child feeding, and Afikepo nutrition programme. At the community level, care groups have been shown to work well and reach most of the mothers in hard-to-reach areas with life-saving messages about nutrition and health.

(4) Large-scale interventions such as the Agriculture Sector Wide Approach Project II, Agriculture Infrastructure and Youth in Agribusiness Project, Agriculture Productivity Programme for Southern Africa, the Sustainable Agriculture Production Programme, and the flagship Social Cash Transfer Programme, which is targeting 10% of the ultra-poor households in all the 28 districts of the country, and the Investing in Early Years for Growth and Productivity in Malawi.

(5) An enabling policy environment with the SDGs at the top of the country's development agenda.

(6) Financial support by the UN and other development partners to the work of different agencies and partners to advance the SDGs.

1.1.6. Gaps and Challenges in the Implementation of SDG2 Activities

(1) Inadequate frontline staff for extension services, which was made worse by Covid-19 regarding limited contact, has been a major challenge in the sector and resulted in low adoption of improved and sustainable agriculture production technologies.

(2) Climate change remains a major challenge to the attainment of SDG2, as most agricultural activities continue to rely on rain-fed agriculture. Natural disasters, especially the recent cyclones, have shown the real potential to erode the gains made. Malawi has suffered from the devastating effects of Cyclone Idai of 15 March 2020, Cyclone Anna in February 2022, Cyclone Gombe in March 2022, and Cyclone Freddy from 12 to 15 March 2023. Frequent cyclones, climate variability, and catastrophes of this kind mean that people experience yearly crop failures and new disasters before they

can recover from the effects of the previous ones, thereby sliding back to poverty.

(3) Poverty, unemployment, rapid population growth, lack of appropriate research-based technologies, innovations and infrastructure, and environmental degradation.

(4) The Covid-19 pandemic, Russia's invasion of Ukraine, and global recession are additional factors.

1.2. Leadership Role of the Malawi Government in Achieving SDG2

The GoM has demonstrated leadership and commitment to achieve SDG2 by providing policy direction, finances, and human resources. There are several ministerial and sectoral-specific policies and strategies to guide the implementation of the country's development and achievement of zero hunger. The Malawi Government additionally takes a leadership role in mobilising public institutions, budgets, and policies in the efforts to achieve its development objectives. It created the Department of Nutrition, HIV and AIDS (DNHA) in the Office of the President and Cabinet in 2004 to provide oversight functions of the national nutrition response, policy and technical guidance, and high-level advocacy for food and nutrition policy and programmes (GoM, 2018). The DNHA coordinates food and nutrition activities with all sectors and ministries, including ministries responsible for health, agriculture, education, finance, and gender among others.

1.3. Roles of Development Partners in Supporting Efforts to Achieve SDG2

Development partners play a key role in supporting governments to accelerate efforts in achieving the SDGs, particularly in developing countries (Eweje et al., 2021). Development partners provide technical and financial support on policy formulation and reviews, development of food, and nutrition programmes and provide expertise in programme monitoring and evaluation. The Malawi

Government recognises the important roles that development partners play in achieving SDG2 and were included in the institutional arrangements for nutrition coordination. In Malawi, development partners supporting implementation of programmes and projects in the food and nutrition sector are organised in an umbrella group called the Donor Nutrition Security Group. This group is responsible for deepening dialogue, coordination, and cooperation and the development of a common approach among development partners on issues of nutrition security.

2. UNIVERSITIES' ROLES IN ACCELERATING THE ACHIEVEMENT OF SDG2 IN MALAWI

Universities play important roles in training a critical mass of graduates, conducting operational research, and developing innovations to accelerate the achievement of SDGs. They are key agents in the education of current and future leaders that will contribute to the successful UN SDGs implementation (Žalėnienė & Pereira, 2021). Education is a significant tool to help reshape individual behaviours, making it crucial for sustainable development efforts. For instance, educated people have higher adaptive capacities and resilience to climate change, while educated women are better equipped to choose safe health options and to protect themselves from trafficking and sexual exploitation (Feinstein & Mach, 2020). Thus, it is indisputable that education plays a crucial role in driving the SDGs given that it can help people transform themselves, their families, their communities, their nations, and the world at large (Abera, 2023) – ultimately, fighting poverty, promoting healthy lives, and ensuring environmental protection and sustainability (Matos & Silvestre, 2013; Mawonde & Togo, 2019). Unlike the Millennium Development Goals that focused on universal primary education attainment, the SDGs incorporate tertiary education into the global development agenda (Owens, 2017). Education at the university/tertiary level has a unique potential to accelerate the capacity to implement the SDGs, given that higher education institutions (HEIs) are hubs of knowledge creation, research, innovation, dissemination, application, and collaboration.

A holistic approach to education, addressing social, economic, and environmental dimensions of sustainability, can foster the knowledge and skills necessary to address the challenges outlined in the SDGs (Lozano et al., 2017) and prepare students to become agents of change in their future careers. Thus, one of the primary ways in which HEIs contribute to the SDGs is through the integration of sustainability-related topics (such as poverty eradication, food and nutrition security, gender equality, climate change, and sustainable consumption among others) into their curricula. In addition, due to concerns on research gaps that exist in linking education and healthy livelihoods within specific contexts, a growing body of literature supports the implementation of policies to address malnutrition often hampered and otherwise influenced by the political economy regarding the distribution of power and resources across society. Thus, discussing multiple ways in which food systems address different forms of malnutrition – food security, food safety, and healthy nutrition – and the tensions associated with power/equity issues among different actor priorities or contexts using a holistic approach is very critical (Balarajan & Reich, 2016; Firdaus et al., 2019; Walls et al., 2019).

There are six public universities in Malawi, namely, Kamuzu University of Health Sciences, LUANAR, Malawi University of Business and Applied Sciences, Malawi University of Science and Technology, Mzuzu University, and University of Malawi (UNIMA). LUANAR was established by an Act of Parliament Number 22 of 2011 by de-linking Bunda College of Agriculture from the UNIMA and integrating it with Natural Resources College. Some of LUANAR's academic programmes that are directly related to SGD 2 include food science, human nutrition, crop science, animal science, aquaculture, veterinary science, agriculture extension, agricultural economics, natural resources management, irrigation engineering, and agriculture education. The contributions LUANAR has made towards the achievement of SDG2, for which it has the mandate and jurisdiction in Malawi, are explored here.

3. LUANAR'S CONTRIBUTION TO THE ATTAINMENT OF SDG2 IN MALAWI

LUANAR conducts training, research, and outreach activities that contribute to improving nutrition using varied genetic resources and plant-based diets in Malawi. These efforts are achieved through research programmes and projects implemented in collaboration with the Malawi Government, DNHA, the Department of Agricultural Research Services under the Ministry of Agriculture, and other research institutions particularly the CIGIAR centres. Key centres that LUANAR works with are the International Crop Research Institute for the Semi-Arid Tropics, the International Institute of Tropical Agriculture (IITA), and the International Potato Centre. LUANAR also collaborates with the University of Nottingham, the Norwegian University of Life Sciences (NMBU), Michigan State University, Arizona State University, and the University of Georgia (UGA). In Africa, LUANAR collaborates with Makerere University, the University of Nairobi, the University of Pretoria, and Stellenbosch University. The research programmes are funded by the Malawi Government and development partners, including USAID, the Department for International Development, the World Bank, the European Union, the Royal Norwegian Embassy, and the UN agencies. These agencies include the UN Development Programme, World Food Programme, and UN Children's Fund. The research collaborations strengthen the research capacity of the faculty at LUANAR and foster innovation, equality and sustainability in health, nutrition, and education in the country.

3.1. Improving the Quality and Safety of Locally Processed Plant-Based Products in Malawi

LUANAR has contributed to the attainment of SDG2 in Malawi through (1) improvements in the quality and safety of peanuts and peanut products; (2) crop biofortification; (3) agro-processing, value addition, and business incubation; (4) promotion of underutilised crops to improve food security; (5) contributing to equality and sustainability through collaborative research on crops;

(6) research, innovation, and knowledge transfer; (7) development of complementary foods; (8) nutrition education on the Malawi six food groups; (9) development of the Malawi Food Composition Database; (10) implementation of projects to foster capacity building related to diets and health, (11) supporting the government in the SUN agenda, and (12) transforming food systems.

3.1.1. Strengthening Productivity of the Peanut Value Chain and Consumption of Peanut-Based Diets

LUANAR participated in the Peanut and Mycotoxin Innovation Lab project led by the UGA to improve the quality and safety of peanut products in Malawi 2012 and 2017. This was part of a broader project implemented in Ghana, Zambia, Malawi, Mozambique, and Haiti under the Feed the Future Innovation Lab for Peanut with support from USAID. The project supported students to study various components in the peanut value chain and provide practical solutions on peanut quality and safety leading to their award of doctoral and master's qualifications (Gama & Adhikari, 2022; Jere et al., 2020). The programme also contributed to equality by encouraging and supporting both male and female students to pursue graduates' studies as well as providing training to female peanut processors on various food peanut-processing techniques so that entrepreneurs can take their products safely to the market. These products include peanut butter, peanut flour, and other foods in which peanuts are important ingredients. The work also culminated into the development of various standards by the Malawi Bureau of Standards, which in a way has scaled up the marketing and consumption of safe products from peanuts in the country.

3.1.2. Crop Biofortification and Geonutrition Research Programmes for Public Health Nutrition

Biofortification improves the micronutrient content of crops. In rural Malawi, where micronutrient deficiencies are very common, biofortification is a promising alternative to supplementation and food fortification programmes since some communities are hard to reach. Biofortification interventions that act through agriculture are likely to play an important role in improving nutrition among many rural and marginalised groups. LUANAR partners with the International Center for Tropical Agriculture and the Africa Research in Sustainable Intensification for the Next Generation (Africa RISING) in developing and testing biofortified beans for improved nutrition and health. The partnership and research are creating opportunities for smallholder farmers and vulnerable categories of people, including women and children, to move out of hunger and poverty. LUANAR also partnered with other universities to study the potential effectiveness of biofortification strategies to alleviate zinc and selenium deficiencies in a research programme called GeoNutrition. The project has provided new evidence to strategies for addressing micronutrient deficiencies in the wider region (Gashu et al., 2021). It has supported several doctoral-level students to study in Malawi and Ethiopia, answering specific research questions on crop biofortification for improved nutrition and health (Botoman et al., 2023).

3.1.3. Promotion of Underutilised Legumes Through Processing and Industry Partnerships

Legumes are some of the nutritious food crops despite being neglected and underutilised in Malawi. Bambara groundnut (*Vigna subterranea*), soybean (*Glycine max*), and cowpeas (*Vigna unguiculata*) are some of the potential legumes that can significantly contribute to nutrition, health equality, and economic sustainability if they are promoted through processing and commercialisation. Bambara groundnut faces another challenge with cultural beliefs that hinder its domestic and commercial utilisation in Malawi (Forsythe et al., 2015). LUANAR has been proactive in promoting these underutilised legumes by conducting research to address the cultural beliefs and taboos as well as processing and industry partnerships to improve farmer adoption, household utilisation, and industrial applications. The university achieved remarkable strides in promoting Bambara groundnuts in a project titled Bambara Market Growth and Nutrition in Malawi, Mozambique, and Tanzania from 2014 to 2018 funded by McKnight Foundation Collaborative Crop Research.

In addition, LUANAR partnered with IITA at Chitedze Research Station on a project that integrated smallholder farmers into the market economy through the soybean value chain from 2013 to 2014. Under this project, LUANAR investigated the knowledge, attitude, and practices of soybean processing and utilisation among smallholder farming families with the aim of understanding dynamics to promote safe soy utilisation in local diets. Besides this, LUANAR has been instrumental in evaluating soy hydrothermal processing methods to reduce anti-nutritional factors and improve bioavailability of nutrients in soy and common bean-based diets (Kalumbi et al., 2019; Kasapila et al., 2024; Mtimuni et al., 2023).

3.1.4. Contributing to Gender Equality and Sustainability Through Collaborative Research on Crops

LUANAR is actively involved in a project on harnessing gender-inclusive research to improve cow peas in Africa, with funding from Feed the Future. In the project, lead scientists from LUANAR have conducted a gender-sensitive value chain analysis to understand the dynamics on how men and women develop preferences around grain yield, taste, colour, pest resistance, and cooking time of different legume varieties. Breeding objectives for new crop varieties often include resistance to drought, low soil fertility, diseases, and pests. By better understanding the needs and wants of local farmers and consumers through the gender-sensitive value chain analysis, the project provided useful information for breeders to develop new varieties of cowpeas. The project has in this regard increased adoption rates of cowpeas in the country.

Another research project on gender equity, equality, and women's empowerment in agricultural production and resource utilisation by LUANAR focused on the relationship between gender roles, legume production, utilisation, and child feeding practices in rural Malawi (Mulenga et al., 2021). This research showed that women farmers are more knowledgeable about legumes and, as such, play an important role in seed selection, storage, and processing. The study further revealed that there is an opportunity to increase women's income by involving them in market information and decisions about legumes.

3.1.5. Sustaining Nutrition Gains Through the Development of Complementary Foods

Malnutrition in Malawi worsens around the first six months because, at this age, the supply of energy and some nutrients from breast milk is no longer adequate to meet an infant's needs (GoM, 2015). A variety of complementary foods is used to supplement the energy and nutrient supply from breast milk. However, these foods are often less nutritious and energy-dense in most of the communities, resulting in acute and chronic undernutrition in young children (Oladiran & Emmambux, 2022). LUANAR champions research on developing nutritious foods from a variety of plant- and animal-based foods to complement breast milk. The research includes testing cultural and sensory acceptability of the foods among caregivers and children in addition to examining potential pathways to commercialise them. Table 5.2 provides a summary of research conducted at LUANAR on the subject matter.

3.1.6. Developing the Malawi Food Composition Database

The GoM supported the development of the Malawian Food Composition Database (MFCDB) and publication of the first edition in 2019. The work was coordinated through the DNHA with support from USAID's Feed the Future Innovation Lab for Nutrition at the Friedman School of Nutrition Science and Policy, Tufts University. South African Food Data System under the South African Medical Research Council and LUANAR provided technical support to the work. A dedicated programme known as the Malawi Food Data System (MaFoods) was established at LUANAR to provide long-term technical support and research for the MFCDB. The aim of the MFCT database is to ensure adequate nutrition and improve standards of locally processed foods through generation and provision of nutrient information. The MFCT describes the nutritive value of locally produced and imported foods that are available in Malawi. The MFCDB is, therefore, pivotal in nutrition and dietetic practice, food preparation at home, and food labelling to ensure nutritional well-being of the general population (van Graan et al., 2023).

Table 5.2.   Summary of Research Conducted at LUANAR on Improving the Safety, Nutrient Quality, and Acceptability of Complementary Foods.

No.	Title	Major Findings and Nutrition Implications	References
1	Effect of incorporating fish (*Engraulicypris sardella*) and processing on nutritive value, acceptability, and shelf life of corn-soy flour blend	Roasting corn and soybeans and the addition of parboiled fish increased the protein content and acceptability of home-made porridges. Corn–soy blend processed from roasted corn, roasted soy, and fish flour can be used an alternative complementary food	Kaunda et al. (2015)
2	Effect of cowpea flour processing on the chemical properties and acceptability of a novel cowpea blended corn porridge	Novel complementary food made by fortification of whole corn flour with three different forms of processed cow pea flours is acceptable by children and their caregivers in rural Malawi	Ngoma et al. (2018)
3	Nutrient content and acceptability of complementary foods formulated from seasonally available foods in Balaka and Dedza districts	The study was designed to address challenges in availability of nutrient dense complementary foods in different seasons of the year. Six appropriate nutrient dense complementary foods accompanied by training on food choices and food preparation were produced. The formulated complementary were acceptable showing potential to effectively solve the problem of undernutrition among infants and young children in the study areas	Waluza and Mwangwela (2018)

4	Relationship between complementary food Aflatoxin exposure and impaired linear growth in 6- to 59-month-old children in Mchinji and Nkhotakota districts	Forty two percent of urine samples from under-five children showed high levels of aflatoxin above 4 ng/mL. Only 20% of caregivers were knowledgeable of the dangers of aflatoxin to health. The research revealed the need for education and training to raise awareness of caregivers on dangers of mycotoxins to child health and ways to mitigate mycotoxin contamination of complementary foods	Mhango et al. (2017)
5	Nutritional quality of Trials of Improved Practices (TIPS)-based complementary food recipes in Malawi: a case study of Kasungu district	This study was conducted to determine the nutritional quality of complementary foods based on TIPS recipes that are frequently fed to children aged 6–23 months among the households that participated in the improved food security and complementary feeding counselling project. TIPS recipes had high dietary diversity, contain more nutrients, and have high percent energy from fats. The improved recipes were more likely to meet nutritional needs of infants than the whole maize flour porridge	Chalemera and Mwangwela (2018)
6	Utilisation of flour milled from micronised (130°c) cow peas in corn-based traditional Malawian snacks		Katungwe et al. (2017)

(Continued)

Table 5.2. *(Continued)*

No.	Title	Major Findings and Nutrition Implications	References
7	Child feeding practices and factors (risks) associated with provision of complementary foods among mothers of children aged 6–23 months in Dedza district of Central Malawi	Complementary feeding is affected by lack of food diversity at household level, increased work burden of mothers and caregivers which led to reduced feeding frequency and inadequate knowledge of mothers to process and prepare nutritious complementary food. To improve the situation, caregivers should be trained on use of energy and time saving technologies, hygiene practices, food processing, preparation, and appropriate child feeding practices	Geresomo et al. (2017)
8	Effect of incorporating legumes on nutritive value of cassava-based complementary foods	Demonstrated that addition of 30% soy in cassava-based commentary foods is effective in improving the nutrient density of the foods in areas where cassava is the main staple crop in Malawi	Kalimbira et al. (2004)
9	Influence of maturity, smoking, and drying of fresh corn on sensory acceptability and nutritional content of complementary porridges	Corn–soy flour blend processed from fresh corn and hydrothermally treated soy can be potentially used in complementary feeding of children at home and school as an alternative to other traditional corn flours	Saka et al. (2018)
10	Acceptability of foods containing Bambara groundnut among children from Bambara groundnut farming households in Ntchisi district of Malawi	Foods containing Bambara were acceptable among school going children (fritters and steamed bread). However, Bambara groundnut milk was less preferred. Availability of ingredients increased the likelihood of the caregivers to process foods containing Bambara groundnuts at home	Katungwe et al. (2015)

4. CONCLUSION

Universities play pivotal roles in training a critical mass of graduates, conducting operational research, and developing innovations to accelerate the achievement of SDG2 globally. The present chapter takes a case study of LUANAR in Malawi and discusses its specific contribution to improving food security and nutrition using varied genetic resources and plant-based diets. The chapter articulates specific examples of work and research and outreach activities conducted to improve availability, access, and consumption of safe and quality food to reduce undernutrition in Malawi. Government leadership, support from development partners, and collaboration with other universities and the private sector are key to success. The research models, impact, and challenges presented in the chapter have relevance and potential for wider application in the developing world.

REFERENCES

Abera, H. G. (2023). The role of education in achieving the sustainable development goals (SDGs): A global evidence based research article. *International Journal of Social Science and Education Research Studies*, 3(01), 67–81. https://doi.org/10.55677/ijssers/V03I1Y2023-09

Balarajan, Y., & Reich, M. R. (2016). Political economy challenges in nutrition. *Globalization and Health*, 12(1), 1–8. https://doi.org/10.1186/s12992-016-0204-6

Botoman, L., Chimungu, J. G., Bailey, E. H., Munthali, M. W., Ander, E. L., Mossa, A. W., Young, S. D., Broadley, M. R., Lark, R. M., & Nalivata, P. C. (2023). *Assessing the residual benefit of soil-applied zinc on grain zinc nutritional quality of maize grown under contrasting soil types in Malawi* (p. 20230197472). agriRxiv. https://doi.org/10.31220/agriRxiv.2023.00182

Chalemera, J., & Mwangwela, A. M. (2018). *Nutritional quality of trials of improved practices (TIPS)-based complementary food recipes in Malawi: A case study of Kasungu District* [MSc thesis]. Lilongwe, Malawi.

Eweje, G., Sajjad, A., Nath, S. D., & Kobayashi, K. (2021). Multi-stakeholder partnerships: A catalyst to achieve sustainable development goals. *Marketing Intelligence & Planning, 39*(2), 186–212. https://doi.org/10.1108/MIP-04-2020-0135

Feinstein, N. W., & Mach, K. J. (2020). Three roles for education in climate change adaptation. *Climate Policy, 20*(3), 317–322. https://doi.org/10.1080/14693062.2019.1701975

Firdaus, R. R., Senevi Gunaratne, M., Rahmat, S. R., & Kamsi, N. S. (2019). Does climate change only affect food availability? What else matters? *Cogent Food & Agriculture, 5*(1), 1707607. https://doi.org/10.1080/23311932.2019.1707607

Forsythe, L., Nyamanda, M., Mbachi Mwangwela, A., & Bennett, B. (2015). Beliefs, taboos and minor crop value chains: The case of Bambara Groundnut in Malawi. *Food, Culture & Society, 18*(3), 501–517. https://doi.org/10.1080/15528014.2015.1043112

Gama, A. P., & Adhikari, K. (2022). Development and optimization of peanut-based beverages: A Malawian consumer-driven approach. *Foods, 11*(3), 267. https://doi.org/10.3390/foods11030267

Gashu, D., Nalivata, P. C., Amede, T., Ander, E. L., Bailey, E. H., Botoman, L., Chagumaira, C., Gameda, S., Haefele, S. M., Hailu, K., Joy, E. J. M., Kalimbira, A. A., Kumssa, D. B., Lark, R. M., Ligowe, I. S., McGrath, S. P., Milne, A. E., Mossa, A. W., Munthali, M., … Towett, E. K. (2021). The nutritional quality of cereals varies geospatially in Ethiopia and Malawi. *Nature, 594*(7861), 71–76. https://doi.org/10.1038/s41586-021-03559-3

Geresomo, N. C., Kamau-Mbuthia, E., Matofari, J. W., & Mwangwela, A. M. (2017). Child feeding practices and factors (risks) associated with provision of complementary foods among mothers of children aged 6–23 months in Dedza district of Central Malawi. *Journal of Nutritional Ecology and Food Research, 4*(1), 15–22. https://doi.org/10.1166/jnef.2017.1146

Government of Malawi (GoM). (2015). *Multiple indicator cluster survey (MICS)*. Zomba, Malawi.

Government of Malawi (GoM). (2018). *National Multi-Sector Nutrition Policy 2017–2021*. Lilongwe, Malawi.

Government of Malawi (GoM). (2019). *Multiple indicator cluster survey (MICS)*. Zomba, Malawi.

Government of Malawi (GoM). (2021). *Malawi Vision 2063 (MW2063)*. Lilongwe, Malawi.

Government of Malawi (GoM). (2022). *Malawi 2022 voluntary national review report for sustainable development goals (SDGs)*. Lilongwe, Malawi.

Jere, A. D., Mwangwela, A. M., Mlotha, V., Phan, U. T. X., & Adhikari, K. (2020). Acceptability of traditional cooked pumpkin leaves seasoned with peanut flour processed from blanched, deskinned and raw peanuts of different varieties. *Scientific African*, *10*, e00598. https://doi.org/10.1016/j.sciaf.2020.e00598

Kalimbira, A. A., Mtimuni, B. M., & Mtimuni, J. P. (2004). Effect of incorporating legumes on nutritive value of cassava-based complementary foods. *Bunda Journal of Agriculture, Environmental Science and Technology*, *2*(1), 13–21.

Kalumbi, M., Matumba, L., Mtimuni, B., Mwangwela, A., & Gama, A. P. (2019). Hydrothermally treated soybeans can enrich maize stiff porridge (Africa's main staple) without negating sensory acceptability. *Foods*, *8*(12), 650. https://doi.org/10.3390/foods8120650

Kasapila, W., Mwangwela, A. M., Njera, D., Matumba, L., Ng'ong'ola Manani, T., Banda, R., & Nyirenda, L. (2024). Sensory acceptability and nutritional quality of composite bread with added puree from biofortified beans in Malawi. *Legume Science*, *6*, e214. https://doi.org/10.1002/leg3.214

Katungwe, P., Mulwafu, A., & Mwangwela, A. M. (2017). *Utilisation of flour milled from micronized (130°C) cow peas in corn based traditional Malawian snacks* [MSc thesis]. Lilongwe, Malawi.

Katungwe, P., Mwangwela, A., & Geresomo, N. (2015). Dietary adequacy of rural school children among Bambara groundnut growing farmers in Ntchisi district of Malawi. *African Journal of Food, Agriculture, Nutrition and Development*, *15*(1), 9620–9634.

Kaunda, G., Mwangwela, A. M., & Geresomo, N. (2015). *Effect of incorporating fish (Engraulicypris sardella) and processing on nutritive value, acceptability, and shelf life of corn-soy flour blend* [MSc thesis]. Lilongwe, Malawi.

Lozano, R., Merrill, M. Y., Sammalisto, K., Ceulemans, K., & Lozano, F. J. (2017). Connecting competences and pedagogical approaches for sustainable development in higher education: A literature review and framework proposal. *Sustainability*, 9(10), 1889. https://doi.org/10.3390/su9101889

Matos, S., & Silvestre, B. S. (2013). Managing stakeholder relations when developing sustainable business models: The case of the Brazilian energy sector. *Journal of Cleaner Production*, 45, 61–73. https://doi.org/10.1016/j.jclepro.2012.04.023

Mawonde, A., & Togo, M. (2019). Implementation of SDGs at the University of South Africa. *International Journal of Sustainability in Higher Education*, 20(5), 932–950. https://doi.org/10.1108/IJSHE-04-2019-0156

Mhango, B., Mwangwela, A. M., Maleta K., & Seetha A. (2017). *Relationship between complementary food Aflatoxin exposure and impaired linear growth in 6-59 months old Children in Mchinji and Nkhotakota Districts* [MSc thesis]. Lilongwe, Malawi.

Mtimuni, B., Munthali, G. T., Gama, A. P., Chiutsi-Phiri, G., Geresomo, N., Malunga, L. N., & Matumba, L. (2023). Hydrothermally-treated soybean-fortified maize-based nsima (stiff porridge) could contribute towards alleviating seasonal body weight loss in farming communities in sub-Saharan Africa. *Heliyon*, 9(7), e17737. https://doi.org/10.1016/j.heliyon.2023.e17737

Mulenga, H., Mwangwela, A. M., Kampanje-Phiri, J., & Mtimuni, B. (2021). Influence of gendered roles on legume utilization and improved child dietary intake in Malawi. *African Journal of Food, Agriculture, Nutrition and Development*, 21(3), 17764–17786.

National Statistical Office (NSO) [Malawi] and ICF. (2017). *Malawi Demographic and Health Survey 2015–16*. Zomba, Malawi, and Rockville, Maryland, USA.

Ngoma, T. N., Chimimba, U. K., Mwangwela, A. M., Thakwalakwa, C., Maleta, K. M., Manary, M. J., & Trehan, I. (2018). Effect of cowpea flour processing on the chemical properties and acceptability of a novel cowpea blended maize porridge. *PloS One*, 13(7), e0200418. https://doi.org/10.1371/journal.pone.0200418

Oladiran, D. A., & Emmambux, N. M. (2022). Locally available African complementary foods: Nutritional limitations and processing technologies to improve nutritional quality – A review. *Food Reviews International*, *38*(5), 1033–1063. https://doi.org/10.1080/87559129.2020.1762640

Owens, T. L. (2017). Higher education in the sustainable development goals framework. *European Journal of Education*, *52*(4), 414–420. https://doi.org/10.1111/ejed.12237

Sachs, J. D., Schmidt-Traub, G., & Durand-Delacre, D. (2016). Preliminary sustainable development goal (SDG) index and dashboard. *Sustainable Development Solutions Network*, *15*, 24–27 http://unsdsn.org/resources/publications/sdg-index

Saka, L., Kasapila, W., Ng'ong'ola Manani, T. A., & Mlotha, V. (2018). Influence of maturity, smoking, and drying of fresh maize on sensory acceptability and nutritional content of the developed porridges. *Food Science & Nutrition*, *6*(8), 2402–2413. https://doi.org/10.1002/fsn3.838

van Graan, A., Chetty, J., Jumat, M., Masangwi, S., Mwangwela, A., Ausman, L., Marino-Costello, E., & Ghosh, S. (2023). Establishing a sustainable Malawian food composition data system harnessing available limited resources. *Journal of Food Composition and Analysis*, *124*, 105705. https://doi.org/10.1016/j.jfca.2023.105705

Walls, H., Baker, P., Chirwa, E., & Hawkins, B. (2019). Food security, food safety & healthy nutrition: Are they compatible? *Global Food Security*, *21*, 69–71. https://doi.org/10.1016/j.gfs.2019.05.005

Waluza, P., & Mwangwela, A. M. (2018). *Nutrient content and acceptability of complementary foods formulated from seasonally available foods in Balaka and Dedza districts* [MSc thesis]. Lilongwe, Malawi.

World Food Program (WFP). (2020). *Malawi country strategic plan (2019–2023)*. https://www.wfp.org/operations/mw01-malawi-country-strategic-plan-2019-2023

Žalėnienė, I., & Pereira, P. (2021). Higher education for sustainability: A global perspective. *Geography and Sustainability*, *2*(2), 99–106. https://doi.org/10.1016/j.geosus.2021.05.001

CASE STUDY: DIVERSE CHARACTERISTICS OF DAILY DIETARY PRACTICES OF CHINESE URBAN RESIDENTS: THE CASE OF GUANGZHOU

Lin Jiahui and Zeng Guojun

Sun Yat-sen University, China

SDG2.1 advocates 'By 2030, end hunger and ensure access by all people, in particular the poor and people in vulnerable situations, including infants, to safe, nutritious and sufficient food all year round'. However, with the trend of global population growth, the challenge of reducing and achieving the goal of zero hunger and providing people with adequate nutrition remains enormous. In China's social context, the frequent occurrence of food safety incidents constitutes the most urgent and complex challenge to society (Shuru, Longjie et al., 2022). Food safety issues have caused many inconveniences to people's daily dietary life, manifested in the proliferation of deep-processed, industrialised food in the market, resulting in people's difficulty in identifying and obtaining healthy food, and even causing many nutritional and physical health problems (Guojun & Longjie, 2019).

To understand how people choose food in their daily lives to ensure nutritional intake, this study has continuously tracked the daily dietary practices of urban residents in Guangzhou from 2018 to the present. This study mainly adopts the field investigation method in which researchers and interviewees jointly complete the activities of purchasing food ingredients, cooking food, and dining together and carry out in-depth communication in the process. The project, supported by Sun Yat-sen University and Newcastle University (UK), explores how to better combat hunger and achieve sustainability in the global dietary system by tracking the dietary practices of city residents in Guangzhou, China; Rio de Janeiro, Brazil; and Johannesburg, South Africa. The project enabled us to discover that the value of higher education institutions lies in the sharing of knowledge and reflection on practice, and the joint search for better solutions.

Our research has found that people have developed a programme to deal with food safety issues in their daily dietary practices. This case summarises the diverse characteristics that Guangzhou residents show in their daily diets. Firstly, urban residents show an attachment to local foods. A large part of Guangzhou's urban residents are migrants from other places. For many non-local residents, they show a deep attachment to the food of their hometown (Guojun et al., 2022). For example, every time interviewee A goes back to his hometown in Guizhou, he brings a lot of pork, local eggs, and even corn to his home in Guangzhou. There is even a special freezer in his home in Guangzhou filled with ingredients from his hometown thousands of miles away. Interviewee A, who insisted on eating pork from his hometown, said, 'I feel that the meat here in Guangdong is not fragrant, and there is no pork flavour after eating it'. Urban residents respond to the lack of access to healthy and nutritious food in urban life by storing local foods, which corresponds to the SDG2 target of sufficient food all year round.

Secondly, urban residents purchase food based on trust relationships. Since the production, transportation, and retail links of food are relatively closed, it is difficult for consumers to obtain real food information. To ensure that the food they buy is authentic, urban residents will look for relatives or acquaintances in their hometown to buy ingredients or look for familiar vendors in local markets (Shuru, Longjie, & Guojun, 2022; Shuru, Hughes et al., 2022). For example, interviewee B, who has lived locally for three years, took us to a food store she often patronises. She said that she prefers to go to trustworthy stores. She believes that the ingredients purchased from stores she knows are fresher and safer. Many urban residents in our study showed similar characteristics.

They identify and purchase safe food based on relationships of trust, which corresponds to the SDG2 target of access to safe food.

Finally, urban residents prefer foods from specific origins. Under the increasingly refined food production system, many agricultural planting and livestock breeding are also becoming increasingly specialised. Many urban residents believe that only food that comes from its origin is 'authentic' and 'nutritious'. During the household survey of interviewee C, we found that his refrigerator contained a lot of mutton transported from Henan. When we asked why, he said,

> *My dad thinks the mutton he bought in Guangzhou doesn't taste that good. It's not hard enough, chewy enough, and a bit soft. He thinks the mutton sent from his hometown (Xinyang, Henan) makes you imagine the sheep when it was alive. It should be very strong and lively.*

In our survey, most urban residents who moved to Guangzhou from other places showed a similar sentiment towards local food. Urban residents ensure the nutritional value of their food by sourcing food from specific places of origin, which corresponds to the SDG2 target of access to nutritious food.

It can be seen from the above characteristics that urban residents build a safe food environment by storing local food, establishing stable food purchase channels, and obtaining healthy food through familiar relationship networks when coping with food safety crises. By doing so, they can maintain a sustainable diet even if they are exposed to unhealthy diets environment, and their practice encourages national and local governments to optimise food policies and create a safer food environment for the public. This is critical to achieving the target of SDG2.1, which is that people have access to safe, nutritious, and adequate food throughout the year.

REFERENCES

Guojun, Z., Jiahui, L., & Shuru, Z. (2022). Multivariate strategies and identity negotiation of trans-local dietary adaptation of immigrant groups: A case study of Guangzhou City. *Progress in Geography*, *41*(4), 660–669.

Guojun, Z., & Longjie, W. (2019). Sustainable food system: Review and research framework. *Geographical Research*, *38*(8), 2068–2084.

Shuru, Z., Hughes, A., Crang, M., Zeng, G., & Hocknell, S. (2022). Fragmentary embeddedness: Challenges for alternative food networks in Guangzhou, China. *Journal of Rural Studies*, *95*, 382–390.

Shuru, Z., Longjie, W., Guojun, Z. (2022). Local food beyond geographical boundaries: The everyday food practices of migrants in Guangzhou. *Human Geography*, *4*(37), 39–45.

Shuru, Z., Longjie, W., Yuchen, X., & Guojun, Z. (2022). Scalar logics of sustainable food systems: Case studies in China, Brazil, and South Africa. *Acta Geographica Sinica*, *77*(8), 2097–2112.

Part 2

THE GLOBAL NORTH: TEACHING AND LEARNING, GOVERNANCE AND COMMUNITY OUTREACH

6

THE POTENTIAL OF CAMPUS FOOD GARDENS TO ACHIEVE STUDENT FOOD LITERACY AND SECURITY IN AUSTRALIA

Sophia Lin[a], Cathy Sherry[b], Tema Milstein[a], Seema Mihrshahi[b] and Sara Grafenauer[a]

[a]University of New South Wales, Australia
[b]Macquarie University, Australia

ABSTRACT

The chapter highlights the growing phenomenon of hunger in affluent nations among vulnerable groups, such as university students. It draws on the results of two studies on food insecurity in the student body at an Australian university in Sydney. It highlights the need and desire of students for increased food literacy at a formative stage of their lives, noting the absence of food growing skills as a recognised part of current understanding of food literacy. The chapter discusses the way in which urbanisation and modern food systems have created such a profound disconnect between people and food production that it no longer occurs to governments and institutions in the Global North that people could grow their own food. The chapter explores historical and global examples of urban agriculture producing meaningful quantities of supplementary food, particularly in times of crisis. Urban agriculture can augment access to safe and nutritious foods (SDG2.1), increase productivity of small producers through knowledge dissemination (SDG2.3), create resilient agricultural practices, maintain ecosystems (SDG2.4), and genetic diversity of seeds through seed-saving practices (SDG2.5). The chapter concludes with a case study of a campus food garden used to increase student food literacy, providing an exemplar for higher education institutions that want to engage with the aims of SDG2 in the context of their own campus.

Keywords: Food security; food literacy; urban agriculture; nutrition; university students; hunger; campus gardens

INTRODUCTION

Student food insecurity can be a risk factor across global campuses, through the high cost of tuition and housing, coupled with limited opportunities to earn income due to academic commitments. These young adults are also embarking on a new stage of life where they have increasing independence, more exposure to diverse people and ideas, and are less influenced by their families. Adding the effects of poor food literacy results in changes to food choices, often towards more affordable energy-dense, nutrient-poor foods leading to poorer nutrition, impacting well-being, and potentially academic performance.

Universities recognise food insecurity as a major student welfare issue but have traditionally limited interventions to providing emergency food relief, vouchers for basic foods, or temporary fiscal support (Hickey et al., 2023). However, these are not sustainable solutions and are largely restricted to universities in the Global North. Students have identified the need for food literacy skills to be delivered as a part of their university experience; opinions that are echoed by researchers (Classens & Sytsma, 2020) who posit that universities are obligated to provide critical food literacy training to improve future food systems.

Establishing campus food gardens may be one strategy to augment food literacy and alleviate food insecurity. Historically and globally, urban food gardens have a demonstrated capacity to produce meaningful quantities of supplementary food, as well as providing social spaces to build community, develop food literacy skills, and engage people in food systems. University food gardens could perform all these functions for the benefit of staff and students. This chapter combines perspectives from public health, nutrition, history, and law to advocate for the potential of university food gardens to teach food skills to staff and students and provide spaces to discuss and engage in sustainable food system practices which are vital to address SDG targets 2.1 (end hunger), 2.3 (increase agricultural productivity), 2.4 (resilient agricultural practices), and 2.5 (genetic diversity of plants and animals).

FOOD INSECURITY ON A LARGE AUSTRALIAN UNIVERSITY CAMPUS

Food security exists when all people, always, have physical, social, and economic access to sufficient, safe, and nutritious food that meets their dietary needs and food preferences for an active and healthy life (FAO, 2009). It occurs on a spectrum and varies in severity from experiencing anxiety that food will run out (marginal food security), to a reduction of the quality, variety, and amount of food consumed (low food security). At the most severe end of the spectrum, people are regularly going without food at all (very low food security) and experiencing hunger. Food insecurity is a powerful determinant of health through its direct impacts as well as indirect impacts on participation in education, employment, and social connection (Loopstra & Tarasuk, 2013). In university students, food insecurity has many detrimental impacts including poor physical and mental health outcomes, poor academic performance, and an increased risk of chronic disease (Hickey et al., 2023). Therefore, campuses which provide healthy food options may help reduce the risk of diet-related health problems (Savoie-Roskos et al., 2022). Ensuring food security promotes social equality and inclusivity and adds to the overall sense of belonging and camaraderie among students. This is important because students are at an important transition period in their lives, often living away from home for the first time and being newly introduced to financial challenges that make it difficult to afford regular meals. Food insecurity is not a new problem and has existed on Australian and American university campuses for many years (Savoie-Roskos et al., 2022; Whatnall et al., 2019), with estimates of between 25% and 50% of students being food insecure (Baker-Smith et al., 2020; Gallegos et al., 2014; Whatnall et al., 2019), indicating that even in affluent nations SDG2.1 (end hunger) is yet to be met for vulnerable populations.

During the first COVID-19 lockdown in Sydney, Australia, between July and September 2020, Mihrshahi and colleagues (Mihrshahi et al., 2022) conducted a survey to explore the experiences of students at Macquarie University (Wallumattagal campus) where there are 44,000 domestic and international students from over 100 countries. The survey was focused on measuring food insecurity (Bickel et al., 2000), psychological distress (Kessler et al., 2002), and the association between these two outcomes (Mihrshahi et al., 2022) to provide governments and tertiary institutions with insights into the resources and support services required by students if similar disruptions occur in future. A total of 28% of students responded that often or sometimes the food they bought did not last and they could not afford to buy more, and 34% stated they often or sometimes could not afford to buy nutritionally balanced meals. Nearly one quarter of students skipped meals and ate less than they should be eating (23%) because of a lack of money. A total of 14% of students reported being hungry. Aggregating this, 42% were categorised as 'food insecure', with 34% having low food security and 8% of students having very low food security at the most severe end of the spectrum. International students were 10 times more likely to experience food insecurity compared to domestic students (Mihrshahi et al., 2022), which has been hypothesised by other researchers to be due to the restriction of Australian government support to international students (Soans, 2021). Food insecurity and psychological distress outcomes were also significantly associated. When compared according to their food security status, food-insecure students had a significantly higher prevalence of high psychological distress than food-secure students (60% vs. 40%) (Mihrshahi et al., 2022).

In direct response to these findings, the University rolled out a series of short-term programmes including emergency grants, grocery e-vouchers, and a free breakfast programme. At a wider level, local and state governments in partnership with local non-government organisations such as Foodbank and OzHarvest[1] provided thousands of emergency food hampers for students at risk. These were often one-off programmes and discontinued soon after the peak of the pandemic passed. Concerningly, data from a repeat survey in 2022 (Dharmayani et al., 2024) showed that the University was no closer to meeting SDG2.1 (end hunger). The prevalence of food insecurity had not changed, with 41% of students still food insecure. Therefore, due to the sustained nature of the problem, long-term solutions such as campus gardens are necessary to address hunger. Regular surveys to collect data on student food security are necessary for universities to monitor progress and enable a data-driven approach to policy and strategy development.

CORRELATIONS BETWEEN FOOD LITERACY AND FOOD SECURITY

Food insecurity among university students often occurs concurrently with low levels of nutrition education, especially in understanding food sources of types of nutrients and how to maintain healthy body weight (Belogianni et al., 2022). Research conducted with Australian university students points to awareness of nutrition education tools such as Australia's national nutrition guideline. However, few follow the available guidance (Lambert et al., 2019). Low levels of nutrition literacy affect all university students, even those studying a health-related degree (Buxton & Davies, 2013; Matthews et al., 2016), indicating that nutrition education interventions should be provided to students in all programmes of study. Campus food gardens provide an opportunity for cross-disciplinary teaching and learning to strengthen student food and nutrition skills.

The delivery of nutrition education must be linked to understanding food systems to reduce the risk of food insecurity. Food literacy encompasses nutrition literacy (Krause et al., 2018), and there is evidence to show that in university students having adequate levels of nutrition literacy is not associated with food security (Moore et al., 2021). Achieving food literacy requires knowledge and skills to source, read, and evaluate information about food, obtain, store and prepare food, and critically reflect on how personal food choices impact on society (Krause et al., 2018). Individuals who have higher levels of food literacy are better able to use their resources more effectively to reduce food insecurity.

The University of New Mexico established campus community food gardens in 2010. These gardens are at a state flagship university in one of the poorest states in the United States, with multiple food deserts abutting the campus. Although on more than 300 hectares, it took years of effort for student and staff advocates to gain administrative approval to establish the first campus food gardens since a WWII-era 'Victory Garden.' After the two gardens planted in 2010 showed they could be successfully supported by volunteer students and staff, other gardens were requested, with volunteers creating a vertical food garden in the student cafeteria used by the chefs, as well as a garden for a University public health research programme. Staff use the gardens to teach across disciplines with hands-on experiential learning and now take students off campus to bridge food-growing and sustainability knowledge. For instance, in a communication course, students study in the campus gardens and at an urban farm with First Nations growers who teach them how to save seeds, that seeds are our ancestors, have led to wars and continuously feed us and the more-than-human world. As they listen, students learn to grind blue corn into flour to make tortillas and then eat with Elders who recount long-standing regenerative food growing histories. This experience teaches students the complex connections between scientific, historic, and geopolitical concepts embedded in SDG2.5.

Fig. 6.1. (*Continued*)

The New Mexico gardens also seeded public partnerships with wide-reaching sustainability effects, including with US Fish and Wildlife Services to collaboratively reimagine the campus as an urban wildlife corridor and refuge. The partnership funded educational public signage and native pollinator habitat, thus addressing more-than-human as well as human food security. While volunteer food growers once battled groundskeepers over poison-spraying and food plant removal, after growing food literacy across campus, these parties began to work collaboratively, with grounds staff even volunteering to plant the campus' first fruit orchard. Though few universities had sustainability plans when our US example began, these are now implemented globally by university professional staff with extensive knowledge of UN Sustainable Development Goals, who are increasingly ready to incorporate food gardens into plans (Milstein et al., 2023).

Fig. 6.1.   University of New Mexico (USA) Campus Community Garden.

However, because the factors that contribute to food insecurity are multiple, occurring at individual, community, and environmental levels, strategies to address food insecurity must be holistic and multi-pronged. Therefore, to improve food insecurity, there must be a focus on food literacy skills, including education on both nutrition and food systems.

Despite this, food-related education and research tends to focus solely on nutrition and treat nutrition education from the point of purchase rather than at the point of food production. This is evidenced by the large amount of research and public health programmes which focus on helping consumers make healthier choices when shopping for and cooking food (i.e. the tertiary food sector), relative to the number of programmes aimed at improving the public's understanding of food systems (the primary and secondary food sectors). Additionally, nutrition professionals also focus on helping clients make healthier choices in retail spaces, providing guidance on skills in shopping and preparing nutritious meals on a budget. However, they are generally not trained to consider whether the client can grow their own food as a way of accessing healthy food.

Frequently used validated questionnaires used to measure levels of food and nutrition knowledge do not include measurement of understanding of food systems. For example, the General Nutrition Knowledge Questionnaire (GNKQ) (Parmenter & Wardle, 1999) focuses on measuring the population's understanding of the national nutrition guidelines, foods where nutrients can be found, nutrient–disease relationships, and food safety. There are no questions dedicated to food systems or food growing. Many researchers and health administrators still use the GNKQ today despite the cultural shift over the past 25 years since its publication towards a more critical view of where food comes from. We cannot expect students to have food literacy if their teachers do not. Therefore, strengthening food literacy among university students also requires academics to improve their understanding of food systems. University food gardens are ideal spaces to facilitate upskilling of educator food systems knowledge.

Food gardens could also be used as a sustainable intervention to address unaffordable food costs on campus, as emergency hampers and vouchers, the most widely adopted strategies (Hickey et al., 2023), do not address root causes. The current Australian university food environment has shifted from university-owned cafeterias to commercial models where unhealthy foods are more accessible than healthy menu items (Coyle et al., 2023). Qualitative interviews on university food environment preferences with Macquarie University students in 2022, as a follow-up to the 2020 food security survey, revealed preferences for free food and meal deals, reintroduction of a university cafeteria where students can buy nutritious, affordable food, access to more nutrition education, cooking and purchasing skills, and an accessible food garden where students and staff grow food. Significantly, the final suggestion of secure access to land to grow food or teach food-growing skills, which would be consistent with SDG2.3 (increase agricultural productivity), has not been adopted by Australian universities to date to address student food insecurity although many universities have large campuses. For example, Macquarie University sits on

126 hectares of mostly under-utilised land. Immediately prior to being part of the University, the land was used for market gardens which provisioned Sydney with fruits and vegetables. While some campuses in Australian universities have food gardens, they are not a prominent part of campus life or curricula. Campus gardens face a range of obstacles to their creation, including deep-seated convictions about what constitutes academic work, a lack of physical space or willingness to prioritise growing space, cost, liability, and limited staff time (Sherry, 2022). Campus administrators – the gatekeepers to growing space – may fear change or have preconceived ideas about campus aesthetics that food gardens disrupt (Milstein et al., 2023).

In comparison, campus food gardens in the United States are far more common than elsewhere (Hagedorn-Hatfield et al., 2022; Milstein et al., 2023; Sherry, 2022) with research suggesting that they improve food security and, consequently, student well-being and academic performance (DeBate et al., 2021; Raskind et al., 2019). A case study from the chapter author Milstein of a US university is presented in Fig. 6.1 and demonstrates how campus land can be transformed to provide learning opportunities for students and enhance the local community and environment.

SOLVING FOOD INSECURITY AND AUGMENTING NUTRITIONAL KNOWLEDGE WITH URBAN FOOD GARDENS

For almost all of human history, the act of eating starts with the production of food from hunting, gathering, or growing. However, in the Global North and increasingly the Global South, urbanisation – the growth in the proportion of a country's population living in urban rather than rural areas – has radically altered this reality. In 2018, 55% of the world's population lived in urban areas (United Nations Department of Economic and Social Affairs, Population Division, 2018), while in Australia, the country where the authors live, that figure surpasses 90% (Australian Bureau of Statistics, 2016). Living in a city means living on insufficient land to produce all of one's own food.

Not producing food has consequences for food security and food literacy. Urban dwellers are heavily dependent on long supply chains which can be disrupted by pandemics, natural disasters, and global conflict (Alabi & Ngwenyama, 2023). Urbanites who have never grown food or seen it grown can have reduced food literacy (Grubb & Vogl, 2019), sometimes being unaware of basic biological facts about plants or animals that determine our food supply, with flow on effects to informed choices about food. For example, most edible plants produce a single annual harvest. However, consumers whose relationship with food starts in a shop stocked with early, mid, and late harvest varieties, sourced nationally and globally, may struggle to understand this. As a result, they are unable to recognise the food miles and potentially compromised nutritional value of out-of-season produce. More importantly for SDG2, urban populations with no experience of food growing are unlikely to understand the significance of SDG2.3 (agricultural production), 2.4 (sustainable and resilient practices), and 2.5 (genetic diversity of plants and animals), or the impact of their own food choices and their nation's trade and agricultural policies on other countries, the environment and small-scale farmers.

Learning from History

History provides empirical and natural experimental evidence for the benefits of widespread urban agriculture on population food security, health, and nutrition. This evidence, supplemented by examples of urban farming such as the case study described here, demonstrates the importance of reintroducing urban food growing and the food skills necessary to facilitate this in urban populations, including in university students.

The increased food insecurity, loss of knowledge, and profound de-skilling that come with urbanisation were clearer to governments in the 19th century during the initial stages of urbanisation that accompanied the Industrial Revolution. At this time, governments still understood that there were *two* ways for the population to source food – purchase with earned income or production with land – and that any disruption to the former could be alleviated by the latter. Long before SDG2.3, governments in the United

Kingdom and mainland Europe provided urban workers with land (allotments, *kleingarten*, *jardin familial*) as a form of poor relief and a means of retaining food-growing skills (Nilsen, 2014; Thorpe, 1969).

When understanding of the benefits of urban food production was beginning to wane at the turn of the 20th century, governments received a brutal reminder in WWI (Thorpe, 1969, p. 16). Imperialism had made Britain dependent on long supply lines and by 1914 much food was imported (Nilsen, 2014, p. 44). The government turned to urban gardeners and allotment holders to help feed the nation. An appeal was made to private gardeners, who had the knowledge to save seeds which were then distributed to allotment holders. By 1917, parks, commons, and playing fields, along with the grounds of schools, hospitals, and other institutions were converted to vegetable and fruit production (Nilsen, 2014, p. 45).

The benefits of an urban population with access to land and the knowledge to use it were demonstrated again with the onset of the Depression, followed by another world war. With the experience of WWI to draw on, the British government was quicker to activate urban food security measures, passing the *Cultivation of Lands (Allotments) Order* in 1939 and launching the 'Dig for Victory' campaign (Thorpe, 1969, p. 19). By 1942, there were over 1,400,000 allotments, with the Thorpe Report concluding that

> [...] *it would be difficult to overestimate the contributions which the produce of allotments made to the nation's food supply during the war years. On 15th March 1944, the Government estimated that food grown ... totalled 10 per cent of all food produced in this country (Thorpe, 1969, p. 19).*

Australian cities have never had an allotment tradition for the simple reason that the iconic Australian quarter acre block gave people food growing space in their backyard. Prior to the 1950s, 'people grew their own food because they *had* to – fruit, vegetables, eggs and perhaps also milk were simply too expensive for families living on modest wages' (Gaynor, 2006, p. 11). During WWII, the Australian government instituted its own 'Grow Your Own' campaign so that urban food growers could make up for shortfalls in food supply caused by diversion of commercially grown food to the armed services and Britain. Food growing takes skill, so part of the campaign was educational. Advertisements encouraged experienced growers to 'help the fellow just starting out ... Those hints of yours ... may well make the difference between the success or failure of his vegetable garden!' (Gaynor, 2006, p. 113). Artificial fertilisers were impossible to obtain. The Haber Bosch process was used to manufacture munitions, and so, the government exhorted people to fertilise by keeping chickens for manure and recycling waste: 'Bury the kitchen rubbish. Never burn a leaf. Anything that will decay will readily add to the humus content of your average garden soil' (Gaynor, 2006, p. 119).

In the post-War period, Australian backyards were often repurposed for pleasure, but food growing continued in the gardens of migrants accustomed to using land productively in their countries of origin. Migrants from Italy and Greece, and later Vietnam, Laos, Cambodia, and China, took advantage of the benign Australian climate and opportunity to own their own home, growing abundant produce that not only nourished their families but also reinforced 'a strong sense of cultural identity and a sense of belonging' (Gaynor, 2006, p. 139).

Other examples of urban food production exist around the world, where, through necessity, city dwellers grow significant quantities of supplementary food. Cuba was forced to revolutionise its domestic food production after the fall of the Soviet bloc and the continued US embargo cut supplies of artificial fertilisers and diesel needed for commercial agriculture (Gonzalez, 2003). The return to urban agriculture and active modes of transport was followed by population-wide reductions in weight, diabetes burden, and cardiovascular mortality (Franco et al., 2013). While cities can never be self-sufficient, historical and global research demonstrates that access to urban growing space can result in meaningful food production and supply, improved health outcomes, as well as increased food literacy. Therefore, campus food gardens should be incorporated into university curricula to address both food security and food literacy problems in the student body, as well as helping to meet SDG2.

INTEGRATING CAMPUS URBAN GARDENS INTO FOOD SKILLS CURRICULAR: AN AUSTRALIAN EXAMPLE

The University of New South Wales (UNSW) Teaching and Research Garden was established in 2019 by an interdisciplinary network of academics (including the first three authors) working on food and wanting to incorporate practical growing into their teaching and research. The establishment of the garden is described more fully elsewhere (Sherry, 2022). UNSW enrols more than 63,000 students from 120 countries. Its Kensington campus cannot expand horizontally due to existing urban infrastructure, resulting in a high-density campus with minimal open space. Despite this, the University was willing to dedicate land for the food garden. A disused concrete slab, next to a multi-story car park, was transformed into a productive food growing space with seven raised beds. Additional growing spaces were created on roof terraces in other campus buildings (UNSW, 2023).

The gardens have been used to increase student food literacy in existing curricula, across disciplines. For example, the gardens are used by the Faculty of Law & Justice in a food law elective to ensure students understood the basic processes of plant reproduction that underpin the national and international intellectual property regimes that facilitate the global, commercial exploitation of plants and seeds (SDG2.4 resilient agricultural practices).

Direct experience of growing food gives students a practical context in which to situate curricula on agricultural, water or trade policies, as well as giving them a tiny experience of the challenges and risks that farmers face from soil degradation and climate change (SDG2.4). Gardens have also been used in land use electives in which students write research essays on urban agriculture, and its potential to green cities and make them more food secure (SDG2.1 and 2.3). The gardens are used by the Faculty of Medicine & Health to teach future dietitians about food production and sustainability, and the role of the SDGs in strengthening future food systems. The teaching staff firmly believe that dietitians who understand where food comes from, and the factors that influence food availability, are vital in helping them to become more compassionate and competent clinicians. This is supported by the national accreditation organisation that has embedded food systems sustainability as a competency standard (Dietitians Australia, 2021). Understanding influences on food availability and price allows them to provide clinical care that is tailored for all populations, including those with reduced incomes.

The UNSW gardens also provide opportunities for the University to engage with the local community. For instance, the gardens are a field trip site for a local secondary school seeking guidance in setting up their own kitchen gardens on school grounds which integrate aspects of the health, science, and geography curriculum. Campus gardens can also provide authentic learning opportunities for university students to develop transferable skills. For example, dietetic students working with local primary and secondary schools to modify school canteen menu items to be healthier (SDG2.2), environmental management students to develop their science communication skills (SDG2.3–2.5), or students developing entrepreneurial skills by creating small business ideas which support sustainable agriculture (SDG2.3–2.4).

During the pandemic, garden leaders contacted student welfare organisations offering garden produce to any international students excluded from government assistance (Soans, 2021), but as Sydney was placed in extended lockdowns in 2020 and 2021, it is not known how many students took up the offer. While the garden is too modest to meet the needs of large numbers of students, the skills it teaches could be implemented in students' backyards and balconies, with home food production supplementing diets as it has done in Australia and overseas at other times of crisis as described earlier.

RECOMMENDATIONS AND CONCLUSION

This chapter highlights food insecurity as a problem in affluent nations for vulnerable populations such as university students, making SDG2.1 a relevant target for tertiary institutions in the Global North. Attempts by universities to alleviate food insecurity have been *ad hoc* and short-term, failing to address underlying causes. The chapter establishes the link between food insecurity and food literacy, demonstrating that increased food literacy for all university students will augment their food security. However, the chapter highlights a missing element in food literacy research as well as in university responses to student hunger – urban food growing. The authors argue that urban agriculture can give urban populations knowledge that is essential not just for their own food choices but for understanding and support of SDG2.3–2.5. Furthermore, history and current global practice clearly demonstrate the capacity of urban agriculture to supply meaningful quantities of supplementary food. These dual benefits of urban agriculture are currently being overlooked by governments and institutions, including universities.

The chapter describes successful campus food gardens in Australia and the United States that have been used in a wide range of disciplines. These exemplar gardens could be replicated on other campuses if administrators are persuaded of the benefits of campus food growing. Campus food gardens offer an outdoor pedagogical environment to introduce students to the socioeconomic, environmental, geopolitical, and cultural complexities of food systems that underpin SDG2. Learning through tending food counters assumptions that such activities are the responsibility of others and make knowledge of growing a legitimate part of higher education. Campus gardens exemplify a transformation-focused approach to education, appropriate for students at a formative stage of their lives. The food garden, as an 'inside-out classroom' (Milstein et al., 2017), fosters collective transformation by starting *inside* from learners' self-identified problems and passions (e.g. hunger, food justice), relating these inner drives to sustainability concepts, and supporting students to apply these concepts *out* in their immediate environment.

The authors argue that providing staff and students with secure access to land, in accordance with SDG2.3, will potentially reduce student food insecurity through provision of supplemental food and development of food skills. Furthermore, by using that land in academic curricula, student food literacy, including understanding of food systems, can be considerably enhanced. SDG2.4 and 2.5 are more likely to be met if populations in the Global North understand how their own food choices, and their nation's agricultural and trade policies, along with geopolitics and legal regimes, affect small-scale food producers, soil quality, ecosystems, genetic diversity, and the world's climate. Universities that are serious about teaching SDG2, as well as meeting its targets for their own populations, would do well to support and fund campus gardens.

NOTE

1. Foodbank is Australia's largest food relief organisation; OzHarvest is Australia's largest food rescue organisation.

REFERENCES

Alabi, M. O., & Ngwenyama, O. (2023). Food security and disruptions of the global food supply chains during COVID-19: Building smarter food supply chains for post COVID-19 era. *British Food Journal, 125*(1), 167–185. https://doi.org/10.1108/BFJ-03-2021-0333

Australian Bureau of Statistics. (2016). *Australian historical population statistics.* https://www.abs.gov.au/statistics/people/population/historical-population/2016

Baker-Smith, C., Coca, V., Goldrick-Rab, S., Looker, E., Richardson, B., & Williams, T. (2020). *#RealCollege 2020: Five years of evidence on campus basic needs insecurity.* https://tacc.org/sites/default/files/documents/2020-02/2019_realcollege_survey_report.pdf

Belogianni, K., Ooms, A., Lykou, A., & Moir, H. (2022). Nutrition knowledge among university students in the UK: A cross-sectional study. *Public Health Nutrition, 25*(10), 2834–2841. https://doi.org/10.1017/S1368980021004754

Bickel, G., Nord, M., Price, C., Hamilton, W., & Cook, J. (2000). *Guide to measuring household food security, Revised 2000.* U.S. Department of Agriculture, Food and Nutrition Service, Alexandria VA.

Buxton, C., & Davies, A. (2013). Nutritional knowledge levels of nursing students in a tertiary institution: Lessons for curriculum planning. *Nurse Education in Practice, 13*, 355–360. https://doi.org/10.1016/j.nepr.2012.09.014

Classens, M., & Sytsma, E. (2020). Student food literacy, critical food systems pedagogy, and the responsibility of postsecondary institutions. *Canadian Food Studies*, 7(1), 8–19. https://doi.org/10.15353/cfs-rcea.v7i1.370

Coyle, D. H., Sanavio, L., Barrett, E., Huang, L., Law, K. K., Nanayakkara, P., Hodgson, J. M., O'Connell, M., Meggitt, B., Tsai, C., Pettigrew, S., & Wu, J. H. (2023). A cross-sectional evaluation of the food environment at an Australian University campus. *Nutrients*, 15(7), 1623. https://doi.org/10.3390/nu15071623

DeBate, R., Himmelgreen, D., Gupton, J., & Heuer, J. N. (2021). Food insecurity, well-being, and academic success among college students: Implications for post COVID-19 pandemic programming. *Ecology of Food and Nutrition*, 60(5), 564–579. https://doi.org/10.1080/03670244.2021.1954511

Dharmayani, P. N. A., Williams, M., Lopes, C. V. A., Ronto, R., Chau, J. Y., Partridge, S. R., & Mihrshahi, S. (2024). Exploring reasons for high levels of food insecurity and low fruit and vegetable consumption among university students post-COVID-19. *Appetite*, 200, 107534. https://doi.org/10.1016/j.appet.2024.107534

Dietitians Australia. (2021). *National competency standards for dietitians in Australia*. Dietitians Australia.

Food and Agriculture Organization. (2009). *Declaration of the world summit on food security*. FAO.

Franco, M., Bilal, U., Ordunez, P., Benet, M., Morejon, A., Caballero, B., Kennelly, J. F., & Cooper, R. S. (2013). Population-wide weight loss and regain in relation to diabetes burden and cardiovascular mortality in Cuba 1980-2010: Repeated cross sectional surveys and ecological comparison of secular trends. *British Medical Journal*, 346, f1515. https://doi.org/10.1136/bmj.f1515

Gallegos, D., Ramsey, R., & Ong, K. W. (2014). Food insecurity: Is it an issue among tertiary students? *Higher Education*, 67, 497–510. https://doi.org/10.1007/s10734-013-9656-2

Gaynor, A. (2006). *Harvest of the suburbs: An environmental history of growing food in Australian cities*. University of Western Australia Press.

Gonzalez, C. G. (2003). Seasons of resistance: Sustainable agriculture and food security in Cuba. *Tulane Environmental Law Journal, 16*, 685–732.

Grubb, M., & Vogl, C. R. (2019). Understanding food literacy in urban gardeners: A case study of the twin cities, Minnesota. *Sustainability, 11*(13), 3617. https://doi.org/10.3390/su11133617

Hagedorn-Hatfield, R. L., Richards, R., Qamar, Z., Hood, L. B., Landry, M. J., Savoie-Roskos, M. R., Vogelzang, J. L., Machado, S. S., OoNorasak, K., Cuite, C. L., Heying, E., Patton-Lopez, M. M., & Snelling, A. M. (2022). Campus-based programmes to address food insecurity vary in leadership, funding and evaluation strategies. *Nutrition Bulletin, 47*(3), 322–332. https://doi.org/10.1111/nbu.12570

Hickey, A., Brown, O., & Fiagbor, R. (2023). Campus-based interventions and strategies to address college students with food insecurity: A systematic review. *Journal of Hunger & Environmental Nutrition, 18*(1), 81–95. https://doi.org/10.1080/19320248.2022.2101413

Kessler, R. C., Andrews, G., Colpe, L. J., Hiripi, E., Mroczek, D. K., Normand, S. L., Walters, E. E., & Zaslavsky, A. M. (2002). Short screening scales to monitor population prevalences and trends in non-specific psychological distress. *Psychological Medicine, 32*, 959–976. https://doi.org/10.1017/S0033291702006074

Krause, C., Sommerhalder, K., Beer-Borst, S., & Abel, T. (2018). Just a subtle difference? Findings from a systematic review on definitions of nutrition literacy and food literacy. *Health Promotion International, 33*(3), 378–389. https://doi.org/10.1093/heapro/daw084

Lambert, M., Chivers, P., & Farringdon, F. (2019). In their own words: A qualitative study exploring influences on the food choices of university students. *Health Promotion Journal of Australia, 30*(1), 66–75. https://doi.org/10.1002/hpja.180

Loopstra, R., & Tarasuk, V. (2013). What does increasing severity of food insecurity indicate for food insecure families? Relationships between severity of food insecurity and indicators of material hardship and constrained food purchasing. *Journal of Hunger & Environmental Nutrition, 8*, 337–349. https://doi.org/10.1080/19320248.2013.817960

Matthews, J. I., Doerr, L., & Dworatzek, P. D. N. (2016). University students intend to eat better but lack coping self-efficacy and knowledge of dietary recommendations. *Journal of Nutrition Education and Behavior*, *48*, 12.e1–19.e1. https://doi.org/10.1016/j.jneb.2015.08.005

Mihrshahi, S., Dharmayani, P. N. A., Amin, J., Bhatti, A., Chau, J.Y., Ronto, R., Turnip, D., & Taylor, M. (2022). Higher prevalence of food insecurity and psychological distress among international university students during the COVID-19 pandemic: An Australian perspective. *International Journal of Environmental Research and Public Health*, *19*, 14101. https://doi.org/10.3390/ijerph192114101

Milstein, T., Alhinai, M., Castro, J., Griegon, S., Hoffmann, J., Parks, M. M., Siebert, M., & Thomas, M. (2017). Breathing life into learning: Ecocultural pedagogy and the inside-out classroom. In T. Milstein, M. Pileggi, & E. Morgan (Eds.), *Environmental communication pedagogy and practice* (pp. 45–61). Routledge. https://doi.org/10.4324/9781315562148-1

Milstein, T., Sherry, C., Carr, J., & Seibert, M. (2023). 'Got to get ourselves back to the garden': Sustainability transformations and the power of positive environmental communication. *Journal of Environmental Planning and Management*, *67*(9), 2116–2134. https://doi.org/10.1080/09640568.2023.2197140

Moore, C. E., Davis, K. E., & Wang, W. (2021). Low food security present on college campuses despite high nutrition literacy. *Journal of Hunger & Environmental Nutrition*, *16*(5), 611–627. https://doi.org/10.1080/19320248.2020.1790460

Nilsen, M. (2014). *The working man's green space: Allotment gardens in England, France, and Germany: 1870–1919*. University of Virginia Press.

Parmenter, K., & Wardle, J. (1999). Development of a general nutrition knowledge questionnaire for adults. *European Journal of Clinical Nutrition*, *53*(4), 298–308. https://doi.org/10.1038/sj.ejcn.1600726

Raskind, I., Haardörfer, R., & Berg, C. (2019). Food insecurity, psychosocial health and academic performance among college and university students in Georgia, USA. *Public Health Nutrition*, *22*(3), 476–485. https://doi.org/10.1017/S1368980018003439

Savoie-Roskos, M. R., Hood, L. B., Hagedorn-Hatfield, R. L., Landry, M. J., Patton-López, M. M., Richards R., Vogelzang, J. L., Qamar, Z., OoNorasak, K., & Mann, G. (2022). Creating a culture that supports food security and health equity at higher education institutions. *Public Health Nutrition*, 26(3), 1–7. https://doi.org/10.1017/ S1368980022002294

Sherry, C. (2022). Learning from the dirt: Initiating university food gardens as a cross disciplinary teaching tool. *Journal of Outdoor and Environmental Education*, 25(2), 199–217. https://doi.org/10.1007/ s42322-022-00100-6

Soans, A. (2021). Pandemic plight of international students will leave Australia poorer. *The Interpreter*. htttps://www.lowyinstitute.org/ the-interpreter/pandemic-plight-international-students-will-leave-australia-poorer

Thorpe, H. (1969). *Departmental committee of inquiry into allotments: Report*. Her Majesty's Stationery Office.

United Nations Department of Economic and Social Affairs, Population Division. (2018). *The world's cities in 2018 – Data booklet (ST/ESA/ SER.A/417)*. https://www.un.org/development/desa/pd/content/ worlds-cities-2018-data-booklet

UNSW. (2023). *UNSW urban growers (UUG)*. https://www.sustainability. unsw.edu.au/get-involved/unsw-urban-growers-uug

Whatnall, M. C., Hutchesson, M. J., & Patterson, A. J. (2019). Predictors of food insecurity among Australian university students: A cross-sectional study. *International Journal of Environmental Research and Public Health*, 17, 60. https://doi.org/10.3390/ijerph17010060

7

FOSTERING STUDENT LEADERSHIP: AN INTERNATIONAL STUDENT CHALLENGE TO ADDRESS SDG2 ZERO HUNGER

Karen Oberer[a], Jolynn Shoemaker[b] and Thomas Rosen-Molina[b]

[a]McGill University, Canada
[b]University of California, Davis, USA

ABSTRACT

In 2022, the SDG2-Zero Hunger Consortium, under the umbrella of the University Global Coalition (UGC), launched an international student challenge. More than 200 students applied to participate from around the world. Eleven teams completed the challenge in May 2023; each team produced a three-minute video explaining an innovative idea for addressing hunger. The first-place team included students from Indonesia, Brazil, the United Kingdom, Costa Rica, Spain, and the United States. Their idea was to establish an End-Hunger Community Center (EHCC) in Indonesia. This chapter describes why and how the SDG2 Consortium developed this challenge and includes a case study from the winning student team.

The chapter provides other higher education (HE) institutions with ideas for engaging students in innovation for the SDGs.

Keywords: Global; competition; student; university; hunger; leadership

INTRODUCTION

There has been much discussion over the past decade about how universities are best able to contribute to the implementation of the United Nations (UN) Sustainable Development Goals. This chapter provides an example of how university students may be engaged on an international level to work collaboratively on developing solutions to SDG2-Zero Hunger. The SDG2-Zero Hunger Consortium, a working group composed of University Global Coalition (UGC) members, developed its International Student Challenge as a means for university students from diverse backgrounds to collaborate on solutions for SDG targets 2.1 (global hunger), 2.2 (malnutrition), and 2.4 (sustainable agricultural practices). The chapter also presents a case study written by the winning team of the student challenge, proposing a novel solution to intersecting hunger and nutrition issues in Indonesia.

The UGC was established in 2019 as a global platform of HE organisations committed to working together and in partnership with the UN Institute for Training and Research, the Sustainable Development Solutions Network, and other relevant organisations, in support of the SDGs. Institutions signed an agreement to support student education and competence in the SDGs, create actionable SDG-related research, and engage with private and public actors to support SDG implementation, among other actions.

The SDG2-Zero Hunger Consortium began in Fall 2021, with EARTH University and the University of California, Davis, as co-chairing institutions. Additional partner organisations joined, including McGill University (Canada), the University of São Paolo (Brazil), and Newcastle University (UK). Soon the Consortium developed key principles: (1) HE institutions play a unique role in addressing SDG2-Zero Hunger; (2) North–South dialogue and collaboration is important, and the consortium's efforts should focus on the inclusion of all members; (3) the Consortium's initiatives should focus on addressing the needs of underserved populations; and (4) the Consortium needs to support the next generation with leadership and interdisciplinary skills to achieve zero hunger.

WHY ORGANISE AN INTERNATIONAL STUDENT CHALLENGE?

In their discussion of educational paradigms for teaching sustainable development, Weber et al. (2021) argued that education has become a neoliberal enterprise where emotions, values, and ethics have generally become devalued in the learning experience, at the expense of workforce training. One of the ways that educators can mitigate this effect is by 'supporting learners to develop their potential in a joint manner for collective reshaping of the future' (Weber et al., 2021, p. 30). The SDGs, which incorporate normative values, are an ideal framework for educators to address feelings of disconnectedness and isolation that sometimes occur when HE curricula are too focused on skill development alone. In the middle of a global pandemic, the urgency to create opportunities for student connectedness was even greater than before.

DEVELOPING THE INTERNATIONAL CHALLENGE ON ZERO HUNGER

The following section details the way in which the planning group originally conceptualised the student challenge; this initial plan was adapted as the project progressed. The subsequent sections illustrate how the challenge unfolded over time (Fig. 7.1).

In 2022, the SDG2-Zero Hunger Consortium began to discuss the idea of hosting an international student challenge. The Consortium members were interested in finding ways to support student leadership on zero hunger and to encourage innovative ideas and direct involvement from young people around the world. The Consortium members agreed that this would be an opportunity to support student leadership, knowledge, skills, teamwork, and international collaboration experience, while also seeding new ideas that students might be able to take forward.

SDG 2: ZERO HUNGER
STUDENT CHALLENGE

1 — CHALLENGE LAUNCHED — NOVEMBER 2022
Challenge launched; marketed internally within universities and externally through the UGC

2 — TEAMS ESTABLISHED — DECEMBER 2022
Consortium staff processed more than 200 applications and established geographically-diverse teams based on research interests

3 — PROPOSALS DEVELOPED — JANUARY 2023
Teams submitted short proposal descriptions; some students dropped out and replacements were found

4 — PROGRESS REPORTS — FEBRUARY 2023
Progress reports due; consortium staff researched and provided resources to the teams according to their proposal topic

5 — OFFICE HOURS — FEBRUARY-MARCH 2023
Office hours scheduled between teams and consortium faculty members to assist teams

6 — SDG A&A WEEK EVENT — MARCH 2023
Three teams invited to speak at Students Taking on Hunger: Ideas from Around the World on Addressing Food Insecurity and Inadequate Nutrition as part of SDG Action and Awareness Week

7 — FINAL GUIDELINES — APRIL 2023
Final submission guidelines shared with team

8 — FINAL SUBMISSIONS — MAY 2023
Final videos/reports due from teams; judges selected from faculty representatives of consortium universities

9 — JUDGING COMPLETED — MAY 2023
All teams informed of the results; every team invited to complete final survey; all receive UGC certificate and letter from UGC president

10 — NEXT STEPS — MAY-JUNE 2023
All teams invited to participate in UGC 17 Rooms; top three teams invited to UGC webinar

Fig. 7.1. International Student Challenge Timeline.
Source: Authors.

In Summer 2022, a planning group from the Consortium began meeting to develop the parameters for the challenge. The subgroup developed a draft concept paper with a description of objectives, structure, and process. The planning group discussed how to structure the teams to support student learning about international collaboration methods. The group determined that four to five students on each team would be optimal so that all could participate fully. There was also consideration about geographic and institutional representation in each team. The group weighed different constructs but ultimately determined that teams with students representing a mix of institutions and countries would be most beneficial for the collaboration experience and would enable diverse perspectives to be represented. Finally, the planning group decided that the challenge should be open to both undergraduate and graduate students, and teams could include both, which had the added benefit of promoting peer mentoring.

The Consortium deliberated about how to address SDG targets and indicators in the challenge. Initially, there was a conversation about anchoring the challenge on specific targets. However, this was determined to be too prescriptive for the purposes of this initiative and for the learning objectives for the students. The SDG targets and indicators do not reflect the full scope of the issues that are inherent in solving SDG2-Zero Hunger, nor do they reflect the intersectional nature of the SDGs. The Consortium wanted to ensure students had the opportunity to consider the challenge of hunger from multiple dimensions and needs, beyond the governmental and multilateral institution lenses, and to recognise connections and trade-offs with multiple SDGs in addressing hunger in specific contexts. However, the Consortium wanted to support student awareness and application of the targets when appropriate. It was decided that the student teams would be required to specify the contributions that their ideas could make to one or more targets and indicators in their final submissions. This ensured that students examined the targets and indicators and thought through the connections to their approaches.

It was also important to leverage the expertise on SDG2 within the Consortium. This was approached through two mechanisms: the Consortium would provide 'office hours' starting at the mid-point of the challenge. The Consortium also decided to collect reading materials on the relevant topics for each team, based on the teams' proposals. The customised reading materials would be made available to each team through a shared folder.

The Consortium also discussed various final outputs for the challenge, ranging from presentations to reports. There was enthusiasm by the Consortium members about a short video output (three minutes) with the Pecha Kucha approach serving as inspiration. The student teams would need to convey the problem and to pitch their idea and the benefits in a concise manner, with latitude for diverse and creative approaches. In addition, two short supplementary reports were added to the structure – one at the mid-point (not to be evaluated on content but to assess progress), and a final written submission at the end to collect more detail on the key elements and input on the challenge experience.

SCOPE AND SCALE OF INTERNATIONAL INTEREST

Realising the Challenge

Once the student challenge was conceptualised, the planning group launched the call for proposals. The following section illustrates how the project evolved as it was rolled out.

Stage 1 – Applying for the challenge: The planning group set up a web form for students to apply to participate in the challenge. The form asked for contact information and university, academic focus, interest areas within SDG2, and motivations for applying.

Stage 2 – Student selection and team assignments: The planning group received over 200 applications from students around the world, much more than expected. Students highlighted some common reasons for applying, including gaining interdisciplinary experience on the issue of hunger and the SDGs; working collaboratively in cross-cultural and international teams; developing practical and creative approaches to a global problem; and having a collective impact beyond what they could do individually. Many students also noted their own personal experiences or community challenges with hunger as an important reason for commitment to this issue.

The high level of interest required the Consortium to increase the number of students permitted in each team, so students were placed in teams of seven instead of four to five. The planning group needed to balance the high level of student interest with the impracticality of placing all 200 applicants in teams. The Consortium was not able to accept all the applicants due to limited infrastructure and capacity; 20 teams of 7 students each were formed. Students were assigned to teams based on students' degree programme, area of interest, and motivations for applying. An effort was made to organise student teams by area of interest, while diversifying according to student level (i.e. graduate or undergraduate), field of study, and geographical location. Some students withdrew from the challenge in the early months, and additional students were selected from those waitlisted to fill the open slots.

Stage 3 – Team agreements and project proposals: The Consortium requested that each team collaborate on a Team Agreement, a type of 'charter' to help determine expectations of one another, preferred means of communication, shared values, and personal development goals. The team agreement encouraged members to reflect on their own personal values and expectations. It provided a structure to reach consensus on team values, desired outcomes, and behaviours. The team agreement guided the participants towards identifying team skills and strengths, determining how to address planning, communications, decision-making, and confidentiality among the team and how to anticipate challenges.

Next, the teams were asked to provide a brief proposal for the project – one-to-two paragraphs describing the issue and their tentative solution. The teams were also provided with guidance about the challenge process at this time, including expected outcomes, helpful tips, expected deliverables and due dates, and information on how the final submissions would be judged.

Stage 4 – Expert resources: The teams were provided with reference materials, based on their proposal topics, and opportunities to join virtual office hours with experts from the Consortium. At each of these office hours, one-to-two Consortium experts answered questions and provided feedback and guidance on the topics and approaches. Consortium administrators also addressed questions about submissions and other steps in the challenge. Office hours were scheduled at different times to account for time zone differences. The fact that the teams were comprised of students residing in different time zones helped ensure that there was at least one student on the team that could attend the office hours. In addition, the Consortium offered office hours upon request. The office hours were not as widely used as the Consortium had hoped, but the teams that did utilise them gained valuable insights that were advantageous in addressing gaps and refining their final submissions.

One of the benefits of organising the student challenge through the Consortium and not through an individual university is that the student teams had access to a variety of experts looking at different dimensions of SDG2 from different disciplines. Moreover, the office hours enabled experts to encourage the teams to closely examine the country-specific contexts for their projects. They helped the teams identify opportunities and partners available in these contexts to address hunger issues.

Stage 5 – Final submissions: Guidance was provided to all teams to help them prepare final submissions. The final report was submitted via an online form which included questions about the roles of each team member; the SDG2 problem and suggested approach/ solutions; other related SDGs; relevance to policymaking at national and international levels; contributions to the Leave No One Behind promise of the 2030 Agenda; logistical challenges and how the team overcame them; and plans to take further action. The teams were asked about valuable aspects of the challenge and areas for improvement for future challenges. The final video submission had a time limit but was open to student creativity and unique team approaches.

Stage 6 – Judging of final projects: Judges for the project were drawn from the SDG2 Consortium members and students from Consortium institutions. Of the 20 groups that were initially formed, 11 completed the challenge. Each judge evaluated the projects based on a rubric (Table 7.1).

Table 7.1. Judging Rubric for the SDG-2 International Student Challenge.

Criteria	UC Davis	EARTH	Newcastle	McGill	São Paulo
• *Knowledge and understanding of SDG2 (20 marks)*					
- Knowledge of SDG2-Zero Hunger					
- Interconnections among SDG2 and other SDGs					
- Consideration of the principle of No One Left Behind					
• *Creative thinking about a problem (20 marks)*					
- Creative idea or solution					
- Originality and creative thinking throughout the process					
• *Appropriate scope (15 marks)*					
- Well-defined scope & recommendations that are feasible to implement					
• *Understanding of context, stakeholders and partners (10 marks)*					
- Teams have explored the specific context for project					
- Understanding of stakeholders and partners					

(Continued)

Table 7.1. (*Continued*)

Criteria	UC Davis	EARTH	Newcastle	McGill	São Paulo
• *Strong communication skills and clear evidence (20 marks)*					
- Clear, science-based evidence to support ideas					
- Easy for layperson to understand the problem and creative idea					
• *Teamwork and collaboration (15 marks)*					
- Commitment and enthusiasm for project					
- Evidence that all team members participated					
Total marks/100					
Additional comments					

Source: Chapter authors Karen Oberer, Jolynn Shoemaker, Tom Rosen-Molina.

The submitted projects were creative and disparate in their approaches. They included the design of an offshore aeroponic farming platform (SDG2.4 sustainable farming practices), a taxation policy aimed at reducing consumption of ultra-processed foods (SDG2.2 malnutrition), a digital platform for sharing climate-smart agricultural practices (SDG2.4), and a project to connect policymakers with vulnerable communities to address hunger (SDG. 2.1 access to food).

Teams were composed of students from universities located in the global North and the global South. The organisers intentionally attempted to match students from different countries; however, they did not know if they were domestic or international students. The organisers noticed that several of the proposed projects were situated in the global South, in countries such as Pakistan, India, Indonesia, Brazil, and South Sudan; two were based in the United States and Canada; and four did not name a specific project location. It appears teams selected the location of their case study based on the expertise and nationality of specific members.

The first-place team proposed the creation of an 'End-Hunger Community Center', which would focus on improving local food systems and nutrition education to address issues of limited accessibility of healthy, nutritious foods in Indonesia. Below is the case study written by this team, which included students from the Universidad Carlos III de Madrid; Carnegie Mellon University; EARTH University; University of California, Davis; and Universidade de São Paulo. Their submission is indicative of the creativity which all students brought to the competition.

Winning Student Team Case Study

END-HUNGER COMMUNITY CENTRE: A COLLABORATIVE STUDENT IDEA TO END HUNGER

Davrina Rianda; Edivan Anjo Ramos;
Hafisat Oladimeji; Privillege Muleya;
Tanaka Murambi; Eviana Barnes

According to the Universal Declaration of Human Rights (1948), everyone has the right to an adequate standard of living, including access to food. Countries have the responsibility to ensure that every person has access to adequate food and that no one gets left behind. However, between 691 million and 783 million people around the globe were hungry, and food insecurity rose to 29.6% in 2022 (World Bank, 2023). As a result, malnutrition is still common, including in Indonesia even though it is Southeast Asia's largest economy and the 10th largest in the world (World Bank, 2022).

Undernourishment has decreased in Indonesia between 2001 and 2020, but other health metrics reveal an alarming nutritional crisis (United Nations in Indonesia, n.d.). As of 2020, 31% of children under the age of 5 suffered from stunted growth; the number of children who were moderately or severely wasted and overweight or obese more than doubled over the past 20 years (United Nations in Indonesia, n.d.). These statistics are concerning for a middle-income country like Indonesia, highlighting the need for more effective efforts to combat hunger.

Our proposed End-Hunger Community Centre (EHCC) aims to combat these issues by addressing not only food scarcity but also the availability of nutritious food, thereby addressing indicators of hunger, such as stunting and obesity that coexist as a double burden of malnutrition within the same family. The hunger issue in Indonesia differs from Sub-Saharan countries such as Somalia and Northern Nigeria, where food insecurity largely stems from conflict. Instead, hunger is linked to limited access to safe and healthy foods, insufficient nutrition knowledge, and poor dietary habits (Paramashanti, 2020; UNICEF Indonesia, n.d.; World Bank, 2015). Driven by a strong community platform observed in Indonesia, we believe that community could act as a catalyst to accelerate zero hunger by 2030. Empowering local communities could lead to more sustainable solutions to end hunger and, subsequently, malnutrition.

PROPOSED PROJECT

We conceive EHCC as a unique community-to-community approach to improve production, distribution, and consumption of foods by focusing on local resources and community connection. EHCC empowers households through a multifaceted approach encompassing three core activities: local food coverage (LFC), healthy eating modules, and home-cook training, along with partnerships for home gardening to distribute locally grown food. Our overarching strategy is to instil practices that can be passed down through generations. Another pivotal aspect of our approach is fostering collaboration between governments and local businesses with local communities to facilitate resource mobilisation and knowledge sharing.

LFC is a term that we created to describe a local food distribution system using food stamps with a personalised approach. It covers not only people living below the poverty line but also those at risk for poverty and food insecurity. What sets LFC apart is its tailored nature, considering the health conditions of beneficiaries assessed by local health-care professionals. LFC is designed to offer multiple levels of

benefits and will be integrated with existing government programmes, including school lunches (Table 7.2). The distribution of food will occur through local markets or home-based
food markets that have participated in EHCC's home-cook
training programme.

Table 7.2. Example of LFC Benefits.

Level	Benefit
LFC Basic	Stamps/tokens/cards for home food supply[a]
LFC Basic Plus	Stamps/tokens/cards for home food supply[a]
	Discounted price for supply for home-based food markets

[a]The benefit will be integrated with existing government programmes.
Source: Student contributors Davrina Rianda, UC Davis; Edivan Anjo Ramos, Universidade de São Paulo; Hafisat Oladimeji, Universidad Carlos III de Madrid; Privillege Muleya, EARTH University; Tanaka Murambi, EARTH University; Eviana Barnes, Carnegie Mellon University.

To support the nutritious food provision for LFC, we propose secondary activities, which are healthy eating modules
and home-cooks training. Our approach begins with a thor-
ough needs assessment and research into local food culture,
ingredients, preferences, and health concerns. The healthy
eating modules will cover vital topics such as the importance
of a balanced diet, portion control, food labels, and healthy
cooking methods. Simultaneously, our home-cook training
programme will serve as the official means to cultivate home-
based food markets that supply LFC. This training encompasses essential elements of food safety, nutrition, meal
planning, and cooking techniques. Participants will engage in
cooking demonstrations and hands-on activities to reinforce
these concepts. The training's primary goal is to ensure that
the foods provided are not only safe and nutritious but also
reflective of local food cultures and ingredients, thus contributing to agricultural sustainability. We will collaborate

with local NGOs, community centres, and health clinics to identify potential participants and suitable training session locations.

At the heart of EHCC is the empowerment of households, and we complement our activities with a home gardening programme. This initiative aims to enhance household access to diverse, nutrient-rich foods and promote the consumption of locally grown food for LFC. Recognising the challenges posed by unpredictable weather patterns due to climate change, we are committed to incorporating modern farming technology and a circular economy system into our community centre. Through cone gardening, a space-saving above-ground cultivation method, we can maximise area utilisation and reduce resource consumption, including land, water, fertilisers, pesticides, and herbicides. This approach not only promotes sustainable agriculture but also offers quicker yields and earlier harvesting. To further align with sustainability principles, we will introduce a circular economy system, utilising organic compost from cooking leftovers and garden waste. Additionally, we will repurpose recycling materials such as plastic bottles as planting pots, contributing to an eco-friendlier implementation.

IMPACT

EHCC will support several indicators of SDG2, including food security (indicator 2.1.2), reducing malnutrition (2.2.2), and productive agriculture (2.3.1). Firstly, our LFC initiative ensures that not only those living below the poverty line but also individuals at risk of poverty have access to their fundamental right to healthy and nutritious food, thereby promoting food security for all (Target 2.1). By directing funds towards essential food purchases, we ensure that resources are used for their intended purpose. Equally crucial, our LFC programme takes a personalised approach based on health

assessments by local healthcare professionals, addressing underlying health conditions, including malnutrition
(Target 2.2). Lastly, our commitment to empowering home-
based food markets and smallholder farmers enhances the
productivity of local food producers and communities, aligning with SDG Target 2.3. We place a strong emphasis on
equal opportunity for all, with a core commitment towards
developing a circular economy.

STUDENT CHALLENGE PARTICIPANT FEEDBACK

At the end of the challenge, the competition organisers sent a survey to the participants which included the following questions:

- What part(s) of the experience did you find most beneficial?
- What part(s) of the experience did you find most challenging?
- What did you learn during the experience?
- Do you have any feedback for project administrators on how
  the challenge was organised?
- Do you have any other input for organisers of the challenge?

Beneficial Experiences

Most of the responses highlighted the following themes: collaboration, international experience, project management skills, tangible
impact, and positive values. By far, the opportunity to collaborate
with other students in an international competition, where they
were exposed to different social contexts, was the most valued
aspect of the challenge. Students also appreciated the ability to
gain research and project management skills, such as performing
literature reviews, brainstorming, problem-solving, and knowledge
sharing. Several also remarked that it was beneficial to be working
on a tangible project that is driven by the need to better the lives
of other people. They enjoyed being able to focus on a particular

region and socio-economic contexts and determining ways to make an impact in the lives of the disadvantaged. Here are some extracts from the survey:

> *For me, every bit of the experience was quite beneficial and worth holding onto. Having different people from different countries with different ideas solving a single problem is an experience in itself. My favourite part is merging our ideas and agreeing on a specific case study.*

> *The development of group work with people from different places, nationalities and professional background[s] was a great challenge, but it was an opportunity to develop skills that could be useful professionally.*

Challenging Experiences

Students noted the difficulty in connecting to others and working asynchronously across different time zones. Other challenges included maintaining team motivation over four months, dealing with unresponsive team members, and some group attrition. The other major challenge was managing workload; several students noted that schoolwork made it difficult to find time to work on their projects.

Learning Experiences

Many students cited learning about different cultural contexts through their colleagues and developing teamwork skills as their most valuable learning experiences. Some mentioned that they developed their interpersonal and communication skills, while others wrote that they deepened their understanding of SDG2 and the interconnectedness of all the SDGs. One contribution summed up many of these comments:

> *During the experience, I gained valuable insights and knowledge that contributed to my personal growth and understanding of the challenges associated with addressing processed food consumption and promoting public*

*awareness of nutritious food … Overall, the experience
broadened my perspective on the interconnectedness of
food, health, and societal well-being, and equipped me
with the knowledge and skills necessary to contribute to
the promotion of nutritious food and combatting processed food consumption in the future.*

Student Feedback for Organisers

The SDG2-Zero Hunger Consortium learned many lessons as the
project progressed and had to adapt to changing circumstances. The
planning group needed to be nimble and engage in frequent communication to complete each stage of the programme successfully.

The planning group asked the students for feedback on the challenge, and there were several recurring ideas:

- Assemble student teams by time zone to facilitate collaboration.

- Be more involved in team organisation and networking; monitor progress more closely to help with team attrition.

- Provide resources in advance, such as online communications
tools, academic writing guides, script-writing instructions,
project development tips, etc.

- Create more opportunities for synchronous discussion between
teams and organisers, such as an introductory online meeting,
workshops, and webinars.

- Offer incentives for participation, such as funding, prizes, and
opportunities to reach a wider audience, including publishing
the reports.

RECOMMENDATIONS

Taking these suggestions into account, the SDG2-Zero Hunger
Consortium can offer recommendations on process for other universities or networks that may be considering a student challenge
on the SDGs. This student challenge also demonstrates some
approaches that can be used to advance student understanding

of SDG2, the targets and indicators associated with it, and the contextual and stakeholder considerations in developing solutions to hunger in different communities around the world. The insights from this process may also benefit universities that are developing student project-based learning on other SDGs and global topics.

Pre-planning

- Ensure adequate lead time to develop all aspects of the challenge and to plan timeline and steps. Six months lead time is recommended if the effort involves multiple offices or institutions in coordination.

Staff Time and Infrastructure

- Determine staff capacity, roles, and responsibilities. Establishing a standing project management group is useful. If the challenge is a multi-unit or multi-institutional project, establish balanced representation and responsibilities.

- Identify relevant online collaboration platforms that can be used for the challenge, and any IT/technical set up that will be needed for applications and submission of deliverables.

Communications

- Develop a communications plan for the challenge. This includes submitting the call for applications through various universities and networks, communicating with individual student participants and teams, and sharing with the public through websites, newsletters, and other mechanisms. If the organisers are from multiple units or institutions, internal communication for updates and decision-making will also be necessary.

- The communications plan should include a detailed schedule that identifies key dates and notifications.

Substantive Expertise and Guidance

- The involvement of subject matter experts can help student teams to build their understanding of the issues and help them refine their ideas. In this challenge, it was essential to include experts with a variety of focus areas under SDG2-Zero Hunger, including agriculture, nutrition, and food security so that all topics could be adequately supported with guidance. Universities should consider the pool of experts who will be consulted well in advance and have a plan for reaching out and including additional expertise if student topics fall outside of the scope of involved experts.

- Subject matter experts can be included through office hours, webinars, and mentoring sessions, for example. They can also provide materials for background reading on the topics that students are exploring.

- Structured events/meetings with subject matter experts should be planned and scheduled in advance. It is helpful to consider ways to make these resources accessible across time zones.

Deliverables and Evaluation

- The required deliverables from the student teams should be planned in advance. In this case, the Consortium considered how to support learning objectives for students to fully understand SDG2-Zero Hunger within the context of the full 2030 Agenda and its principles.

- Details on how to submit and guidance for deliverables should be provided to student teams as early in the process as possible.

- Student teams should also be informed of the evaluation process at the beginning of the challenge to guide their projects.

Student Recognition

- Although some student challenges offer cash or other prizes to winning teams, this is not always necessary.

- Students should have opportunities for both individual and team recognition. This may include certificates of completion for the challenge, listing participants on organisation websites, and letters of congratulations from the organisation that is sponsoring the challenge and/or the student's university.

- Featuring the winning teams in an in-person or online event is another way to bring attention to the projects and help students gain visibility and possible support for their ideas.

CONCLUSION

Overall, the SDG2 International Student Challenge was a success. One participant commented that the challenge 'was a transformative experience'. Another wrote,

> [t]he challenge was well-organised and well-planned. International students from around the globe were invited, teams were made and reading materials were provided. Time zone challenges impacted the learning process and mentorship facility, however, meeting this challenge was also a learning experience.

The SDG2-Zero Hunger Consortium is developing additional collaborative initiatives, and the experience with the student challenge is directly informing the group's planning for future student learning opportunities on SDG2-Zero Hunger (both curricular and co-curricular). The energy, fresh ideas, and personal commitment to solving hunger that the students brought to this challenge have, in turn, inspired the involved universities to pursue new opportunities to support global student learning across countries and regions on SDG2-Zero Hunger.

The participating students had a strong desire for opportunities to address pressing sustainability challenges in concrete ways. Whipple et al. (2021) observed a similar need in their case study. They write that students often feel the need to 'do something' to address sustainability challenges, but typical science curricula do not equip them appropriately. They write, '[a]s student interests in sustainability increase, it is important to also provide them with the tools needed to pursue careers that contribute to and encourage environmentally targeted solutions while crossing multiple sectors' (Whipple et al., 2021, p. 4). We propose that online competitions such as the SDG-2 International Student Challenge are one of the tools in the toolbox that can assist students in developing solutions to hunger and other complex global challenges of our day.

REFERENCES

Paramashanti, B. A. (2020). Challenges for Indonesia zero hunger agenda in the context of COVID-19 pandemic. *Kesmas: Jurnal Kesehatan Masyarakat Nasional (National Public Health Journal)*, *15*(1), 24–27. https://doi.org/10.21109/kesmas.v15i2.3934

UNICEF Indonesia. (n.d.). *Nutrition*. Retrieved October 9, 2023, from https://www.unicef.org/indonesia/nutrition

United Nations in Indonesia. (n.d.). *Sustainable development goal 2: Zero hunger*. Retrieved October 9, 2023, from https://indonesia.un.org/en/sdgs/2/progress

Universal Declaration of Human Rights. (1948). *GA Res 217A (III), UNGAOR, 3rd Sess, Supp No 13, UN Doc A/810 71*. Retrieved October 9, 2023, from https://www.ohchr.org/en/human-rights/universal-declaration/translations/english

Weber, J. M., Lindenmeyer, C. P., Liò, P., & Lapkin, A. A. (2021). Teaching sustainability as complex systems approach: A sustainable development goals workshop. *International Journal of Sustainability in Higher Education*, *22*(8), 25–41. https://doi.org/10.1108/IJSHE-06-2020-0209

Whipple, S., Tiwari, S., Osborne, T. C., Bowser, G., Green, S. A., Templer, P. H., & Ho, S. S. (2021). Engaging the youth environmental alliance in higher education to achieve the sustainable development goals. *Scholarship and Practice of Undergraduate Research*, 5(1), 4–15.

World Bank. (2015, April 23). *The double burden of malnutrition in Indonesia*. Retrieved March 22, 2023, from https://www.worldbank. org/en/news/feature/2015/04/23/the-double-burden-of-malnutrition-in-indonesia

World Bank. (2022, April 5). *The World Bank in Indonesia*. Retrieved October 9, 2023, from https://www.worldbank.org/en/country/indonesia/ overview

World Bank. (2023). *Food Security | Rising Food Insecurity in 2023*. World Bank. Retrieved October 9, 2023, from https://www.worldbank. org/en/topic/agriculture/brief/food-security-update

8

A COMPILATION OF GLOBAL CASES ON TEACHING, LEARNING, AND CAMPUS STEWARDSHIP

ABSTRACT

The final chapter of this book captures a diverse range of case studies across teaching and learning and campus stewardship. The five case studies presented here add illustrative insights of global actions, following on from the four case studies presented within previous chapters. These contributions provide compelling and practical insights into how Higher Education Institutions (HEIs) are broadly addressing SDG2 through student competitions in the first section, and curriculum approaches in subject areas in which 'food' and 'hunger' might not be expected in the second section. The final section of cases further explores HEIs' approaches to food security through both a curriculum lens and a student-led campus project. It is a fitting way to close the book since the cases collectively illustrate the potential contribution that any student and/or educator might make to raising awareness of and action towards achieving zero hunger and SDG2.

Keywords: Food insecurity on campus; course innovation; student engagement

STUDENT COMPETITIONS

Triggering Change – An SDG2 Challenge Competition
Hosted by John Cabot University in Rome

Michèle Favorite and Silvia Carnini Pulino

John Cabot University, Italy

The Triggering Change (TC) competition at John Cabot University
(JCU) in Rome is a two-minute SDG2-related pitch that students
create after benefiting from expert talks by international entrepreneurs who share their best practices. Students then develop their
own entrepreneurial proposals and present them in creative ways
via a video. Their eco-literacy is measured before and after their
pitches through the Sulitest online survey (Sulitest, undated), which
assesses general SDG and specific SDG2 competencies. In the last
edition, general SDG eco-literacy increased by 22% and SDG2
literacy was up 7%.

JCU believes that diverse backgrounds enrich the collective
discussion on sustainability and that creativity develops a positive mindset. As a US liberal arts university (the largest in Europe),
JCU promotes the pitch across its 14 academic departments and
includes students who are both degree-seeking (from 75 countries)
and visiting (mostly American). TC's original 2022 edition included
participants from four disciplines; in 2023, these had risen to 10:
history, psychology, international affairs, public relations, Spanish,
communications, marketing, economics, finance, and business. To
increase diversity, other universities were invited to join. In 2023,
JCU partnered with Richmond University in London, which contributed 100 MBA students. In total, TC's last edition included 196
students from 80 countries. The following global experts provided
examples of inspirational, innovative solutions to SDG2 throughout the food value chain.

- *Bruno Ferreira, Co-Founder, Banco de Alimentos de Bolivia
  (BAB).* BAB is the largest Bolivian food bank and aims at
  redressing inefficiencies in domestic food systems. Malnutrition
  affects over half of Bolivians (World Food Programme, 2022).

This triggers the 'hunger cycle', which impacts nearly all other SDGs, hindering development and self-fulfilment, and short-circuiting education, health and gender equality. BAB aims at improving Bolivia's food systems at the primary and agricultural levels of the food chain, where lack of training, good practices, investment, accessibility, and technology prevail. Plus, poor roads and storage facilities impact crop productivity and spoil harvested products before they reach retail. Ferreira specifically covered malnutrition (Targets 2.1 and 2.2), agricultural productivity (2.3), food production systems (2.4), and food commodity markets (2.5).

- *Andrea Petrini, Italy Sales Lead, Too Good to Go*, is the global mobile app provider that connects customers to restaurants and stores that have surplus food. He highlighted the developed world's need to improve food systems at the end of the value chain, where most food waste occurs. He described the scale of food waste and its impacts. Forty percent of all food produced is wasted and inevitably creates a domino waste effect: all resources required for food that is unconsumed are used in vain (food systems use 30% of the world's energy, mostly from fossil fuels (Food and Agriculture Organization of the United Nations, International Renewable Energy Agency, 2021)) 8–10% of greenhouse gases are associated with unconsumed food (United Nations Environment Programme, 2021). Petrini explained his organisation's two goals: reduce food waste by restaurants, food stores, and supermarkets by connecting them with consumers; educate consumers about responsible consumption; and offer practical tips (freezing, cooking, using one's senses instead of discarding food that has been mislabelled). This talk explored the interconnectedness among SDGs 2, 12, and 13.

- *Janina Peter, Head of Innovation, World Food Forum (WFF)*, is an arm of the United Nations (UN) Food and Agriculture Organization. She described the WFF's mission as a global youth forum that fosters solutions in food governance and supports entrepreneurial ideas related to SDG Targets 2.1 and 2.4.

The input from these three experts provided students with a compelling vision from which to develop their ideas into an entrepreneurial pitch. Some student testimonials of how this positively impacted them are included here.

1. Isabella Giner, American University (Washington, DC, USA), Spanish Major

> *I focused on sustainable food production systems and agriculture (SDG2.4 and 2.5). Before TC, I didn't know about the negative impact of meat consumption on agriculture, which contributes hugely to greenhouse gases. Children need to know about this, so I created a book for them.*

> *My story focuses on bears living in a magical forest. Their meat consumption leads to the degradation of their once-thriving environment. When they realize this, they embark on a journey to a magical tree for wisdom. It urges them to adopt sustainable agricultural practices and consume less meat. They heed the tree's advice, and their forest starts to heal. The bears' journey represents a metaphor for humanity's relationship with nature. This experience has reinforced my appreciation for the power of storytelling in conveying complex concepts.*

2. Sara Segat, John Cabot University, Communications Major (Co-winner, Fall 2023)

> *Thanks to TC, I learned about SDG2, specifically SDG2.1, and its interrelation with SDG12. I discovered how complex SDG2 is and that food affects almost everything in our lives, both in developed and developing countries. My idea combined two approaches: reducing food waste and donating to food shelters. Even in the developed world, the needy and lonely are victims of malnutrition. I chose an idea that could have real impact in a developed country,*

like Italy. It consisted of a gamified food delivery experience where users could gain points if they agreed to have their takeout meals prepared with slightly imperfect or surplus resources that would otherwise be wasted. This would allow food providers to limit food waste and contribute to a rewards program where users convert credits – earned by choosing sustainable food options – into donations to food shelters. It's a partial solution because, as I have learned, food security requires a multi-dimensional approach. But it's a starting point to mobilize privileged citizens to contribute to food security.

3. Sahari Abney, John Cabot University, Business Administration Major

My idea was a cookbook for university students, the greatest food wasters globally due to poor food organization and decision-making. My goal was to reduce food waste by educating students about the benefits of shopping for ingredients with long shelf lives and using them in tasty recipes. Students need to learn that saving food makes more of it available to those in need. TC taught me that food affects us all, it expanded my horizons and motivated me to enact change.

In conclusion, TC provides a 360° learning opportunity. Students improve their SDG2 knowledge in five weeks (measured via the Sulitest survey) and realise its interconnectedness with many other SDGs. They acknowledged its complexity but are not daunted by it. Through the example of experts, students felt empowered to envision solutions and make their voices heard. They also learned that SDG2 and its ramifications manifest themselves differently in different parts of the world: in Bolivia, with malnutrition; in Italy, through food waste. They understood that complexity can be addressed through creativity and that diversity provides a richness of perspectives in addressing SDG2 through, for example, a children's book or a cookbook.

Love Student Leftovers: A Digital Student Cooking
Competition in the UK and Ireland

*Karen Cripps, Pariyarath Sangeetha Thondre
and Jo Feehily*

Oxford Brookes University, UK

In 2022, Oxford Brookes University launched a digital student cooking competition as part of a regional initiative by the United Kingdom (UK) and Ireland chapter of the 'Principles of Responsible Management Education' (an initiative supported by the United Nations). It formed part of a national network series of events on the theme of food and sustainability. The competition brought together students, academics, and civic partners to raise awareness of food waste, promote responsible consumption, and offer solutions in terms of reducing waste and addressing food poverty in the higher education student community. The project achieved these objectives through the competition, an expert panel discussion of sustainable food systems by civic partners, and through research carried out into hospitality businesses' approaches to food waste and food rescue. The successful template of this project contributed to the success of a 'SDG Challenge' funding grant by the Association of Commonwealth Universities in 2023. This enabled the competition to run again and expanded to cover all university students across the UK and Ireland, in partnership with Amity University (India) to cover all students across India. It serves as a helpful example of partnership for the goals (SDG17) in raising awareness and action around food waste (SDG12) and interconnections with SDG2 by addressing issues of food poverty and nutrition in cooking.

The competition invites students to post video and/or photo series recipes on Instagram, which used food that is commonly wasted and/or uses leftovers. Entries in 2023 fall into three categories; using up 'whole food and vegetables' and commonly wasted foods, 'rescued and repurposed' foods (using surplus foods) and 'throw it-all in' cooking (to encourage using up scraps and creative cooking). The content raises awareness of global challenges such as

Competition Categories

£100 — Whole fruit and veg & commonly wasted foods

£100 — Rescued and repurposed food

£100 — Throw-it-all-in cooking

Create a recipe using entire fruit & vegetables (including the peels that you would normally throw away) and commonly wasted foods and have the chance to win a prize of £100

Reuse your leftovers, use food picked up through food rescue apps or 'yellow sticker cooking' (from the 'reduced' section at the shops) and have the chance to win a prize of £100

You don't need to go shopping again! Get creative and use up what is lurking in the back of your cupboard, freezer and fridge), throw it all in and have the chance to win a prize of £100

Fig. 8.1. Instagram Content from the Competition.

Source: Author (Karen Cripps, Pariyarath Sangeetha Thondre and Jo Feehily).

the paradox between food waste and food hunger (Fig. 8.1). Creative social media content was designed to be informational, concerning why food waste is a problem in terms of wasted resources; aspirational, showing how it is possible to reduce environmental impact while also saving money, and engaging, through recipe and resource ideas.

Recipe ideas enable students to make better use of their budgets, thus addressing food poverty in the student community. The home website for the competition, which included educational material about food waste and cooking tips, nutrition, and competition guidelines, further enhanced the educational reach of the project. Students from the UK and India are communicating and sharing experiences through the project design on the challenges and opportunities surrounding students in addressing food waste, through a global lens.

Much of the project success is attributed to the support of a network of civic partners that represented all aspects of the sustainable food system, from carbon menu planning to cafe menus based on the use of surplus food. The partners acted as judges for the competition and took part in campus events in which they demonstrated cooking using up commonly wasted and/or surplus food items (Fig. 8.2). This allowed all students on campus to learn about initiatives in the local area such as charities that collect surplus food and redistribute it to vulnerable people in the community. This can be surprising to students who are based in a relatively affluent area of the country (Oxfordshire, UK). Associated with the competition and campus events were research projects, including a survey of the approaches of Oxfordshire hospitality businesses to food waste and food rescue solutions, and a behaviour-change survey to identify student attitudes towards food waste and behaviour-change mechanisms.

The project enabled an interdisciplinary approach, with global collaboration, that can potentially unite university students across three countries (UK, Ireland, India). The competition offered a fun, creative, and accessible way to raise awareness about the SDGs. It also provided a compelling context for universities to engage with local civic organisations and communities at the front line of addressing food waste and food poverty.

Fig. 8.2.   Students Interviewing Civic Partners During a Leftover 'Cook-off' Challenge on Campus.

Source: Author (Karen Cripps, Pariyarath Sangeetha Thondre and Jo Feehily).

Student Perspectives

Gracie Ball (BA Hons Business and Management)

When getting involved in this project, I had no idea how much food waste a problem in Oxfordshire was or anywhere. Visiting places such as Oxford Food Hub, Cherwell Collective and Waste2Taste really gave me insight into how big the food waste problem is and how well charities such as these help combat this problem. This made me see the big role/impact that these community kitchens have on our society and how beneficial they are. The connection between food waste and poverty has such a strong correlation and I think if we combat food waste (for example, through competitions such as this one), this will help reduce the amount of food poverty there is in the country which is a huge issue at the moment! I think it's very helpful for students to learn about food waste through competitions such as LoveStudentLeftovers as it makes education on this topic more fun and entertaining and helps give students recipes and other tips to use instead of throwing their food away!

Siana Dimitrova (BSc Hons Nutrition)

During the three years of my course I have lived with various students and a common tendency I have noticed among them was the frequent purchasing of 'ready-meals' and 'take-away' food. However, a lot of these meals do not contain essential food groups such as fruit and vegetables. As a nutrition student, I understand that this is a serious issue, and I am aware that this pattern could stem from a fear of cooking from scratch. This is why when I found out about the idea of the lovestudentleftovers competition, I thought I could make a change among my peers and encourage them to eat healthier. By providing educational content regarding food waste and highlighting the ways home cooking could help, I believe students will be encouraged to explore their culinary skills and include more whole fruit and vegetables in their diet, while also saving money.

CURRICULUM INNOVATIONS

Zero Hunger Training for English Language Teacher Candidates in Turkey

Ilknur Bayram[a] and Özlem Canaran[b]

[a]Turkish National Police Academy, Türkiye
[b]TED University, Türkiye

The teacher education programme in Turkey consists of theoretical and practical aspects of English teaching, but there is no emphasis on the integration of the Sustainable Development Goals (SDGs) into English language teaching (ELT). Research shows that language teachers do not consider sustainable development (SD) as relevant to their field, and they do not know how to address the SDGs in their lessons (Kwee, 2021). This is partly due to the insufficient coverage of SDGs in pre-service education, and/or a lack of on-the-job training opportunities on why SDGs matter in ELT (Bedir, 2021).

Education for Sustainable Development (ESD) promotes the SDGs by (1) embedding them into the curriculum; (2) having all education institutions and stakeholders share the principles and values supporting sustainable living and actions; (3) promoting problem-solving, critical thinking, and action that addresses the sustainability challenges; (4) applying diversity in education methods; (5) involving learners in decision-making; and (6) raising awareness in addressing local and global challenges (Nevin, 2008). Centred around the ESD approach, this case study presents the findings of a project aimed at incorporating SDGs into the ELT curriculum in Turkey. The project was conducted with student teachers, who – among other SDGs – specifically studied SDG2 Zero Hunger. Special attention was paid to SDG2 not only because the UN estimates every nine people worldwide are suffering from hunger (Kılıç, 2022), but also, in Turkey, food waste amounts to 26 million tons annually (Hamzaoğlu & Göktuna, 2022), making SGD2 a global and a national concern.

This study is part of a larger project titled 'Piloting the "Sustainable Development Goals in English Language Teaching" Course Syllabus'. The project was conducted with 13 third- and fourth-year ELT student teachers at a foundation university in Turkey. Student teachers participated in a 14-week online training delivered by a teacher trainer with a PhD and extensive experience in ELT. Researchers developed the course syllabus to provide student teachers with ESD competencies that would assist them in developing SDG-focused lesson plans. The syllabus consisted of activities introducing the history and background of SDGs; teaching approaches in ESD; mapping the SDGs onto the primary, secondary, and tertiary level ELT curriculum and designing instructional materials addressing SDGs.

In week four, student teachers learned about SDG2. This case study was bounded by the work of six student teachers who planned a lesson aligned with SDG2. They designed their plan based on feedback from the trainer and classmates and implemented it in their practicum school (see Fig. 8.3 for the phases of the project). The lesson plans focused on cognitive (e.g. knowing about the causes and effects of hunger on human life), socio-emotional (e.g. feeling empathy and responsibility for people suffering from hunger), and behavioural learning objectives (e.g. evaluating and implementing actions

Training

Focus on Theory
SDG2 Zero Hunger
SDG2 & ELT

Collaborative Lesson Planning

Peer Collaboration
Collaborative lesson planning on SDG2 in ELT

Implementation

Focus on Practice
Implementing the lesson plan in practicum school
Receiving feedback from peers and mentor

Fig. 8.3.   Phases of the Project.

Source: Authors (Ilknur Bayram and Özlem Canaran).

to combat hunger). SDG2 was introduced through videos, songs, news extracts, and information from the UN (United Nations, 2023) and UNESCO (UNESCO, 2023) websites. Practice parts on the plans employed role-plays, scenario, and project development. While the resources provided on the UN and UNESCO websites have offered valuable insights into SDG2 topics like food waste, malnutrition, and food insecurity, there's still a pressing need for more educational materials and classroom activities that cater to learners of varying ages and abilities. These resources would help learners develop the essential competencies linked to SDG2.

The session on SDG2 received positive reactions from student teachers. It started with a discussion of SDG2, its definition and main targets; cognitive, socio-emotional, behavioural learning objectives teachers might develop as part of their lesson plans; and suggested topics and examples of learning approaches for SDG2. Such a progression was reported to have a positive impact on student teachers' understanding of world hunger as a threatening problem and how to make it part of their teaching practices. One participant commented;

> Our trainer's lecture on SDG2, the sample outcomes, activities and lesson plans she shared with us were amazing. Its' being not too theoretical and full of sophisticated concepts was one thing I especially liked. We need to see more practical stuff to understand how SDGs work in action.

Creating a lesson plan as a group enabled student teachers to hold discussions on zero hunger, heightening their awareness of SDG2. They reported benefiting from different viewpoints about how to conceptualise the problem and how to exploit it in class. One participant mentioned;

> I was used to designing lesson plans on my own, which can be very limiting. This experience was unique because I could learn how five other teachers felt about zero hunger, how they would make a lesson plan out of it, and what resources and activities they would make use of for language teaching purposes.

Receiving feedback from the trainer and classmates was also observed to facilitate the lesson planning process.

Student teachers implemented their lesson plan in their practicum school and received feedback from their mentor teacher and the students. This experience influenced their understanding of how SDG2 issues like reducing food waste, mindful eating practices, and ensuring food security could be effectively integrated into teaching. Moreover, activities such as zero hunger challenge games, and creating posters on Zero Hunger played a crucial role in engaging language learners in an authentic classroom setting. One teacher rightly highlighted; 'If we hadn't implemented the plan, it would have been the biggest missing link in the project'.

Training, lesson planning and implementation phases increased student teachers' knowledge of SDG2 and facilitated its integration into their lesson plan, resulting in an observable increase in their self-confidence. For this reason, in countries with insufficient financial resources allocated for teacher education centring around ESD, it might be reasonable to start with such a low-cost and practice-oriented scheme. The case is promising as it presents a practical model of how we can train potential teachers to make them more ESD oriented. Practicum courses in the teacher education programmes serve as a bridge between theoretical knowledge and practical application. These courses can enable teacher candidates to witness firsthand how their SDG2-focused lesson plans translate into real classroom settings, especially when ESD becomes integrated into teacher education curriculum.

Sustainable Food Systems: A Live Event for Accounting Students at a University in Northern Ireland

Xinwu He

Queen's University Belfast, UK

Due to the recent significant shift in stakeholder demands for sustainability reporting, accounting professionals are expected to understand the expanded scope of an accountant's role and equip themselves with adequate learning and development (IFAC, 2024).

One strategy to develop sustainability reporting competence is via holistic learning in the higher education context (Caldana et al., 2023). Moreover, addressing the SDGs from an interdisciplinary perspective can stimulate students' systematic thinking, problem-solving skills, and anticipatory competencies for sustainability-related issues (Alm et al., 2021).

This case study presents an innovative pedagogical design which follows an interdisciplinary approach (Alm et al., 2021) and integrates the knowledge of SDG2 and connected SDGs, sustainable food systems, and sustainability accounting and reporting. The pedagogical design consists of two parts, drawing on the idea of holistic learning (Caldana et al., 2023): (1) an in-class tutorial on *sustainable food system and accounting for the SDGs* and (2) a live learning event on the same topic, combining outdoor activities and informative speeches. After engaging in the learning activities, students are expected to (1) identify and apply SDG good practices in real-life scenarios and (2) critically evaluate the evolving role of accounting and the accounting profession in supporting the SDGs. The following sections discuss the pedagogical design in detail, followed by the Concluding Remarks summarising its benefits, possible challenges, and recommendations.

The tutorial aims to familiarise students with the knowledge of SDG2, other connected SDGs, and sustainability accounting. Firstly, the lecturer explains what a sustainable food system is and why our food choices matter. Students are encouraged to reflect on food-related issues in their daily lives. Secondly, the lecturer introduces how the food system works, covering key concepts including regenerative agriculture (SDG2.4), empowering farmers (SDG2.3), animal welfare (SDG2.4 and SDG12), food distribution (SDG2.4 and SDG12), carbon footprint (SDG2.4 and SDG13), healthy diet (SDG3), and food waste (SDG2 and SDG12).

Thirdly, from an accounting perspective, the lecturer introduces how organisations can identify material sustainable development issues, account for their environmental and social impacts, and disclose relevant information. Taking SPAR as an example, the students are asked to identify how each of its responsible retailing initiatives (disclosed in *SPAR International Annual Review*) contributes to various SDGs, including SDG2. Drawing on this example, the students can reflect further on why the role of accountants is changing and how accounting can improve the way it serves society.

The live event represents a novel mode of delivery, which involves learning activities outside of the classroom. Fig. 8.4 shows the running plan of the event, which consists of four sessions.

By engaging in this live event, students can apply the interdisciplinary theoretical knowledge acquired to business practices. The event implicitly emphasises food waste prevention. With a more sustainable food habit, every individual, as a consumer, (potential) employee, and citizen, has the power to contribute to SDG2. Students reflect on what actions they will prioritise to promote habits that benefit their health and the planet. Their reflections are assessed via an innovative method which can enhance their digital practice experience and learning skills.

The pedagogical design helps accounting students acquire interdisciplinary theoretical and practical knowledge. By linking SDG2 and sustainable food system with accounting and reporting, it stimulates accounting students' passion for challenging the traditional definitions of accounting, reimagining the accounting profession's role in a sustainable future, and thus linking accounting's full potential to their future career development. It also represents an effort to reinvigorate the university accounting curriculum

	9:00 – 12:00 A trip to a local farm based in Northern Ireland.
	12:00 14:00 Sustainable Lunch at Queen's Business School. Staff members from the Campus Food and Drink give a speech about Queen's University Belfast's *Sustainable Hospitality*.
	14:00 – 15:30 Guest speakers from SPAR give a speech about their responsible retailing practices. A visit to a SPAR convenience store.
	15:30 – 17:00 Group Assignment: Podcast Production. Requirement: each group should create a 6-minute podcast to summarise what they have learned during the day and what actions they will prioritise to contribute to SDG 2 and connected SDGs.

Fig. 8.4. Live Event Plan.

Source: Author (Xinwu He).

and encourage students to think about accounting as something beyond a mere technical tool. Moreover, the pedagogical design allows students to take ownership of their learning process (Alm et al., 2021) and generate transferable learning resources (e.g. podcast episodes) that benefit the broader teaching and learning community. Accounting educators are encouraged to try similar pedagogical designs that embed interdisciplinary knowledge, holistic learning approach, and active learning process.

This pedagogical design can be embedded into university accounting (or, after necessary adjustments, business/management) undergraduate or postgraduate modules, suitable for students from diverse cultural backgrounds. Notably, the implementation can be challenging, especially when it comes to the live event. With a large student cohort, organising outdoor learning activities might be difficult, given the cost associated with the farm trip and sustainable lunch. Also, due to health and safety reasons, proper risk evaluation should be done in advance, which can be time-consuming. Meanwhile, implementing the pedagogical design requires extensive support. This means that the module team needs to make extra efforts to coordinate with other departments and external guest speakers.

Where student cohorts are large or time and resources are restricted (e.g. other parts of the world where accounting might be understood and taught differently), lecturers may want to consider bringing guest speakers into the classroom to share their practical experiences and interact with the students. Furthermore, other innovative learning activities can be designed, such as preparing scorecards that account for students' daily food carbon footprint.

Meeting Development Goals in Education: An Interdisciplinary Approach Focused on Food at the University of York, United Kingdom

Tim Doheny Adams, Ulrike Ehgartner and James Stockdale

University of York, UK

The global food system is giving rise to a growing social, health, and environmental crisis with unsustainable practices seen across food production, supply chains, and consumption (Willett et al., 2019). These challenges are highly complex and interrelated and require interdisciplinary cross-sector systems-level analysis, structural and institutional change, and education to solve. The UK's higher education system excels at training specialised skill sets within subjects such as plant biology, consumer psychology, environmental law, and many more to address specific challenges within the food system. But equally crucial is clear communication among future graduates working in and on different sectors of the food system to accurately identify opportunities, risks, and unintended consequences associated with future interventions. Language, values, approach, and motivations can differ dramatically between disciplines and pose a significant barrier to truly integrated understanding of the issues from multiple perspectives. University students rarely can directly develop these interdisciplinary-related skills, in part, because the development of interdisciplinary programmes presents several pedagogical difficulties. Disciplines and teachers may be pedagogically isolated, students' may lack academic vocabulary or methods of inquiry needed for studying a particular subject, and there can be difficulty maintaining a balance between the different interdisciplinary approaches (Yang, 2009).

'Future of food' is one of a suite of sustainability-related interdisciplinary modules recently (2023/2024) rolled out at the University of York open to all students and aims to address this skills and knowledge gap by providing a truly multidisciplinary learning experience that examines the wicked problems in our food system from diverse perspectives. The content of the module is divided into four distinct, but strongly interconnected topic areas, each addressing critical aspects of the food system through different lenses: (i) climate change, (ii) agricultural inputs, (iii) food waste, and (iv) sustainable and healthy diets. Unique to this module is the balanced collaborative approach to teaching, with each topic area being co-developed and co-delivered by academics from different disciplines. For example, the section on agricultural inputs has been developed by a biologist and management scholar and is explored from both a scientific perspective (how plants grow, the environmental impact of fertiliser/pesticide usage, etc.) and a business/management perspective (how farmers manage risk, shocks to the price of agricultural inputs, etc.). The topic on healthy and sustainable diets introduces students to economic dynamics and perspectives on responsibility and accountability of businesses and consumers and is collaboratively taught by a sociologist and a management scholar.

Moreover, contributions to the course came from beyond academia, with practitioners from the food system invited to contribute lectures on various topics, where they share their perspectives, visions, and challenges. For example, a local farmer discussed sustainable, regenerative farming techniques, providing his insight on supporting ecosystems and soil health within the economic constraints of running a rural farming business, and a risk assessments team leader from the food standards agency highlighted some important regulatory frameworks regarding public health protection. This interdisciplinary approach enriches students' understanding by integrating real-world insights with academic knowledge, empowering them to consider complex challenges with practical solutions. By bringing together experts from various fields, the module fostered interdisciplinary dialogue and collaboration, breaking down silos that often hinder comprehensive problem-solving.

The interdisciplinary nature of this undergraduate module reflects ongoing major research collaborations across the University of York and was designed to mirror the interconnectedness of the real-world food system. By integrating insights from social and natural sciences, students should become better equipped to grasp the systemic challenges facing food production and consumption. This approach responded to the need for holistic solutions that address social, environmental, and economic dimensions simultaneously and consider various stakeholder interests and needs and in interest from students in challenge-led research. Furthermore, the module provided the students, who themselves worked in interdisciplinary teams, with valuable opportunities to develop skills relating to the effective communication and collaboration across disciplines. This is the first year the module is running and our modest cohort of 30 students includes students from psychology, environmental science, linguistics, chemistry, and biology. We are expecting more than 100 students to enrol on this module next year.

Institutional support has been key to the development of this module. The University of York strategy 2030 highlights environmental sustainability and interdisciplinarity as two of the main guiding principles towards the realisation of a university for the public good. In 2021, the university launched ESAY (Environmental Sustainability at York), a collaboration between research, teaching, policy, and campus operations across all disciplines and involving external partners. Additionally, in 2023, an institutional restructure of teaching from three terms to two semesters and institutional-wide standardisation of modules to 20 credits opened opportunities for more cross-disciplinary collaborations on teaching. The York Interdisciplinary Modules (Future of Food, Sustainability and Policy, Climate Crisis Action Lab, and Sustainability Clinic) were developed to support the integration of environmental sustainability and interdisciplinary collaborations into existing programmes.

Existing interdisciplinary expertise on the topic of food systems has also been fundamental to the development of the Future of Food module. Three academics teaching on this module are involved in FixOurFood, an interdisciplinary research programme anchored at the University of York and funded by the £47.5 million Transforming UK Food System, Strategic Priorities Fund Programme delivered by UKRI. The programme takes a place-based and co-productive approach to enable food system transformation with a focus on farming, businesses, and school food and includes interdisciplinary research projects involving a diverse set of stakeholders. The team has, therefore, been able to leverage existing contacts, interdisciplinary experience, and focused food system expertise to design holistic and realistic learning activities.

CAMPUS PROJECTS AND CURRICULUM ADDRESSING HUNGER IN THE GLOBAL NORTH

Soul Food: How Two Canadian Students' Legacy Saved 100,000 Pounds of Food

Matt Hopley

Queen's University, Canada

Queen's University, Canada, is home to over 27,000 students, located in the town of Kingston, Ontario (Queen's University, 2021). With a population of approximately 175,000, a recent survey identified the homeless population of Kingston to be over 200 (Statistics Canada, 2022). To care for this vulnerable population, especially during harsh northern winters, many shelters exist in the form of free overnight hospitality, kitchen meal services, and social church programmes. These facilities, however, only attain food in the form of donations, and they frequently run out of supplies (Feed Ontario, 2023).

It is a sad reality that before 2008, the three dining cafeterias on campus (which serve over 350,000 meals per year to first-year residents) were forced to waste over 10,000 pounds of food per year. Campus cafeterias prepare fresh meals for students every day. Cafeterias are spread out across campus and students' meal plans provide them access to all venues. This causes uncertainty about the amount of food chefs must prepare each morning, resulting in a consistent overestimate to avoid running out of food at any one cafeteria. At the end of each day, anything that has been prepared or cooked is thrown away. This fact alone is disheartening; however, worse is the fact that homeless shelters running out of food are less than five kilometres away. It is commonly believed that being one of the most developed countries in the world, food insecurity does not exist in Canada. However, this is far from true. A recent census revealed that 6.9 million Canadians (18% of the population) had experienced some level of food insecurity in the past year (Statistics Canada, 2023). Of this population, 5% were considered severely insecure and reported missing meals, having a consistently reduced food intake, and going multiple days without food

(Statistics Canada, 2023). So, in 2008, with hopes of providing a local solution and bringing the world closer to Sustainable Development Goal (SDG) 2.1: reducing food insecurity, a pair of then third-year students brought together a team of 60 undergraduates and devised a plan with university cafeteria administration.

To this day, the routine is the same. Groups of one to four students commit to a weekly volunteer shift for the duration of the school year. On the same morning each week, they drive in their personal vehicles to one of three campus cafeterias where the staff has packaged last night's leftovers in pans and trays. The students then deliver the food to a specified homeless shelter, drop off the leftovers, pick up the previous delivery's now clean pans and trays, and deliver them back to the cafeterias before completing their approximately 20-minute shift. Taxis are also occasionally provided to volunteers who do not have their own car.

At Queen's University, student-run clubs can become ratified by a yearly student vote, sanctioned by the university. Ratified clubs are then officially recognised by the school and receive a small portion of student ancillary fees. Soul Food receives just over $1,000.00 from the school each year in student fees. This money is used for taxis to shuttle students who volunteer but do not have cars, as well as supplies like food transportation bags, pans, trays, advertisements, volunteer appreciation events, and donations to shelters. Clubs at Queen's function to fill many different purposes, with some solely operating for the benefit of students, and the community, raising awareness, or fundraising for different causes. Soul Food is an exceptional club because it checks all these boxes and, in turn, shines a very positive light on the university and its relationships within the community.

The club is now composed of an executive committee, consisting of two co-chairs and a marketing, finance, social media, website, graphic design, and logistics manager. This team is responsible for the annual fall hiring of volunteers, scheduling of volunteers, promotion of the club, and communication with cafeterias and shelters. Outgoing co-chairs remain involved for a following year, overseeing and providing any help or guidance that may be needed by each year's new committee. Presently, one of the club's primary goals is to help other schools start Soul Food chapters. This is a unique programme that has not been reported to exist anywhere else in the country.

Since founding the club in 2007, over 100,000 pounds of food have been rescued from campus cafeterias and delivered to local homeless shelters in Kingston. The club's volunteers deliver enough food to feed between 50 and 100 people every night. Soul Food continues to have a profound impact on both the shelters and the members being fed, as well as the volunteers. 'Soul Food has been one of our primary food sources during the chilly, winter months here in Kingston' reported one shelter when reflecting on their relationship with the club. 'If not for the students at Queen's who deliver us these donations, many of the individuals who rely on our services would not get fed'. But current Soul Food volunteers also benefit from this club ... 'Soul Food has provided me with a sense of meaning in my community and given me joy in getting to see the smiles of grateful community members who I get to help feed'. Perhaps this is why the club has no trouble recruiting ample volunteers each year. Despite volunteering being an unwritten requirement for Canadian University students wishing to enter the workforce (as employers highly value volunteer work and community engagement), Soul Food is an especially attractive volunteer opportunity because of the fulfilment, social environment, and low time commitment. Requiring no more than 20 minutes of a volunteer's time per week, it takes cafeteria employees even less time to wrap the uneaten food than it would to throw it out. Soul Food is built on a simple, replicable model that is powerfully self-sufficient, surprisingly undemanding, yet immensely impactful.

When Hunger Is in Your Higher Education Classroom in the United States

Xenia K. Morin

Rutgers University - New Brunswick, USA

Sustainable Development Goal (SDG) 2, 'Zero Hunger' is often applied to developing countries but hunger exists in America too. Hunger was well documented in 1985 in the book *Hunger in*

America, authored by the Physicians Task Force (1985). The findings of the wide-reaching study made the national press because the testimony was shocking – hunger and malnutrition were present in America. But little was reported about hunger on our college campuses in 1985.

Hunger, now often referred to as 'food insecurity' (SDG2.1), has not gone away. In 2022, the White House National Strategy called for 'Universities, colleges, and academic medical centers should bolster hunger, nutrition, and physical activity research and data collection disaggregated by factors, including race, ethnicity, and other demographic and social factors' (White House National Strategy on Hunger, Nutrition and Health, 2022, p. 35). What this report recommends, while long overdue and excellent, fails to recognise that there are additional roles that universities and colleges can play: to acknowledge its own students' food insecurity and address it through research, education, and outreach. Fortunately, work is already underway that can help illuminate the issue on college and university campuses.

Hunger on college campuses in the United States can be invisible until it is not. As a faculty member, it is not unusual to have students struggling in class, but when you are informed by the student, that part of the struggle is because they are struggling to find housing and food, that is when it hits you: hunger is in your classroom. Hunger is no longer abstract; it has a human face: a student who makes every effort to sit in front of you, to do the work, and to keep up with class while struggling to meet basic needs. This is the student who, if you gain their trust, tells you there is no food at home and you direct them to the student pantry and other resource and hope that they can remain in school. How do you help this student? How can your classroom, your teaching, give voice to this student while also helping the student to understand that they are not alone?

Recent scholarly work confirms that hunger, food insecurity, and other basic needs requirement are more prevalent in our classrooms than many faculties knew. Rutgers opened its student food pantry in 2016, and in 2018, Rutgers published the results of a survey of food insecurity which found that 'a third of Rutgers students were food insecure, with 36.9% of undergraduate students

and 32.2% of graduate students reporting some level of food insecurity' (Cuite et al., 2018, p. 10). An expanded survey, looking at
basic needs, was reported in 2020. More questions were added to
provide insight on food insecurity and the picture this data painted
was heart wrenching. Data included responses to the phrase 'I was
hungry but didn't eat because there was not enough money for
food'. We learn that during the 30 days before the survey, 20.6% of
undergraduates and 17.7% of graduate students responded positively to this statement (Cuite et al., 2020, p. 10). One ray of hope
was that awareness of the student pantry had risen from 20.1% in
2016 to 66.1% in 2019 in the undergraduate student population.
This change might be due to outreach efforts during new student
orientation along with a request to include the language in course
syllabi that advises students on who to contact in case of difficulty
in accessing sufficient food.

Specific curriculum courses such as 'Introduction to Agricultural
and Food Systems' discuss food insecurity and make it visible.
Others on campus also discuss this topic. Students watch and
comment on films such as Silverbush (2013) and Seidel (2015),
which explore hunger in America's school as well as more locally
produced films, selected because they featured student voices and
talked about the live experience of food insecurity. The assignment
is to watch and discuss the films in class or watch asynchronously and respond via discussion boards so they can respond to one
another. The films allow me to get students to problem solve and
ask students what they would do to address food insecurity if they
lived in the communities featured in the film? Some students bring
their lived experiences to these discussions.

At the beginning of my courses, I ask students to explore why
they eat what they eat. In this exercise, students are given sticky
notes or pieces of paper to brainstorm on the reasons why they eat
what they eat. I give them up to 10 minutes to come up with 15–20
reasons for why they eat what they eat; they write one reason per
sticky note. If students are working online, they can use the stick-
ies in a Jamboard (Google) document. Students work in groups of
three to five to generate ideas. They often stop after two to three
minutes, and encouragement is needed to get them to 10 or more

reasons; but they can do it. Once the group has 15–20 reasons, they are asked to group the reasons and assign them to categories. The purpose of this exercise is to help students realise that access and cost or affordability, along with accessibility and convenience and marketing/social media are key drivers of their food choices. Of course, there is an assumption that there are choices when you ask your students to do this exercise. However, at the end of the exercise, it can be extended to a conversation about food insecurity and its causes. Interestingly, this exercise also lets students know that you can be a resource to discuss food insecurity, access, and affordability.

Alongside other institutional approaches, such as the Rutgers student farm, which is supervised by a staff member, students, through paid internships, learn how to grow food, some of which goes to the university's student pantry. These activities, from seed to consumer, have a direct connection to SDG2 by producing nutritious food. We also teach indoor cultivation and introduce students to hydroponic food production so that those without gardens can produce food. There is much that colleges and universities can do to be a part of the national discussion on food insecurity and SDG2 and faculty and staff can play an important role.

REFERENCES

Alm, K., Melén, M., & Aggestam-Pontoppidan, C. (2021). Advancing SDG competencies in higher education: Exploring an interdisciplinary pedagogical approach. *International Journal of Sustainability in Higher Education*, 22(6), 1450–1466.

Bedir, H. (2021). English language teachers' beliefs and perceptions on sustainability. *International Journal of Curriculum and Instruction*, 13(2), 1880–1895.

Caldana, A., Eustachio, J., Lespinasse Sampaio, B., Gianotto, M., Talarico, A., & Batalhão, A. (2023). A hybrid approach to sustainable development competencies: The role of formal, informal and non-formal learning experiences. *International Journal of Sustainability in Higher Education*, 24(2), 235–258.

Cuite, C. L., Brescia, S. A., Porterfield, V., Weintraub, D. S., & Willson, K. A. (2018). *Working paper on food insecurity among students at Rutgers-New Brunswick.* http://humeco.rutgers.edu/documents_pdf/RU_Student_Food_Insecurity_2018.pdf

Cuite, C. L., Brescia, S. A., Willson, K., Weintraub, D., Walzer, M., & Bates, L. (2020). *Basic needs insecurity among Rutgers-New Brunswick students.* https://go.rutgers.edu/BasicNeedsStudy

Feed Ontario. (2023). *Hunger report: Why Ontarians can't get ahead.* https://feedontario.ca/wp-content/uploads/2023/11/FEED_OntarioHungerReport23.pdf

Food and Agriculture Organization of the United Nations, International Renewable Energy Agency. (2021). *Renewable energy for agri-food systems – Towards the sustainable development goals and the Paris agreement.* https://www.irena.org/-/media/Files/IRENA/Agency/Publication/2021/Nov/IRENA_FAO_renewables_Agrifood_2021.pdf

Hamzaoğlu, N. M., & Göktuna, B. Ö. (2022). Food waste behaviour of organic food consumers in Turkey. *Journal of Management and Economics Research*, 20(4), 209–224.

IFAC. (2024). *A literature review of competencies, educational strategies, and challenges for sustainability reporting and assurance.* International Federation of Accountants (IFAC).

Kılıç, R. (2022). The problem of hunger in the world and a new model proposal to solve this problem. *Balkan Journal of Social Sciences*, 11(21), 63–68.

Kwee, C. T. T. (2021). I want to teach sustainable development in my English classroom: A case study of incorporating sustainable development goals in English teaching. *Sustainability*, 13(8), 4195.

Nevin, E. (2008). Education and sustainable development. *Policy & Practice-A Development Education Review*, 6, 49–62.

Physicians Task Force. (1985). *Hunger in America: The growing epidemic* (1st ed.). Wesleyan University Press.

Queen's University. (2021). *Student affairs annual report, 2020–2021.* https://www.queensu.ca/studentaffairs/sites/vpsawww/files/uploaded_files/Annual%20Reports/DSA%20Annual%20Report%202020-21.pdf

Seidel, D. K. (2015). *Generation at risk: Joining forces to fight childhood obesity*. Rutgers University. https://ifnh.rutgers.edu/generation-at-risk/

Silverbush, L., Jacobson, K., Goldman, J., Harrington, R., Colicchio, T., Bridges, J., Patel, R., & Burnett, T.-B. (2013). *A place at the table*. Magnolia Home Entertainment.

Statistics Canada. (2023, November 14). *Food insecurity among Canadian families*. https://www150.statcan.gc.ca/n1/pub/75-006-x/2023001/article/00013-eng.htm

Statistics Canada. (2022, July 13). *Focus on Geography Series, 2021 Census* – Kingston, Census metropolitan area. https://www12.statcan.gc.ca/census-recensement/2021/as-sa/fogs-spg/page.cfm?lang=E&topic=7&dguid=2021S0503521

Sulitest. (undated). *Mainstreaming sustainability literacy*. https://www.sulitest.org/

The White House. (2022). *Biden-Harris administration: National strategy on hunger, nutrition and health*. https://www.whitehouse.gov/wp-content/uploads/2022/09/White-House-National-Strategy-on-Hunger-Nutrition-and-Health-FINAL.pdf

UNESCO. (2023). *SDG Resources for Educators – Zero hunger*. https://en.unesco.org/themes/education/sdgs/material/02

United Nations. (2023). *Goal 2: Zero hunger*. https://www.un.org/sustainabledevelopment/hunger/

United Nations Environment Programme. (2021). *UNEP food index report*. https://www.unep.org/resources/report/unep-food-waste-index-report-2021

White House National Strategy on Hunger, Nutrition and Health. (2022). https://www.whitehouse.gov/wp-content/uploads/2022/09/White-House-National-Strategy-on-Hunger-Nutrition-and-Health-FINAL.pdf

Willett, W., Rockström, J., Loken, B., Springmann, M., Lang, T., Vermeulen, S., Garnett, T., Tilman, D., DeClerck, F., Wood, A., & Jonell, M. (2019). Food in the Anthropocene: The EAT–Lancet Commission on healthy diets from sustainable food systems. The Lancet, 393(10170), 447–492.

World Food Programme. (2022). *Bolivia (Plurinational State of) annual country report.* https://docs.wfp.org/api/documents/WFP-0000147933/download/

Yang, M. (2009). Making interdisciplinary subjects relevant to students: An interdisciplinary approach. *Teaching in Higher Education, 14*(6), 597–606.

ABOUT THE EDITORS

Karen Cripps is a Senior Lecturer in Responsible Management and Leadership at Oxford Brookes Business School and a Senior Fellow of the Higher Education Academy. With a long-established academic and teaching profile in sustainability leadership and management, her practice is deeply aligned with the United Nations 'Principles of Responsible Management Education' (PRME). She is the Secretary for PRME UK and Ireland Chapter Steering Committee and an Ambassador for PRME Global Sustainability Mindset Working Group. She has led several student and research projects to address SDG2, including funding by the Association of Commonwealth Universities in 2023, under its SDG Challenge Grant.

Pariyarath Sangeetha Thondre is a Senior Lecturer in Nutrition and the Research Lead for Oxford Brookes Centre for Nutrition and Health at Oxford Brookes University. She is a Registered Nutritionist (Association for Nutrition) and a Fellow of the Higher Education Academy. With more than 15 years of academic and research experience, she teaches Food Science and Global Nutrition and Public Health to undergraduate and postgraduate students. She is a member of the 'Education for Sustainability' working group at Oxford Brookes University and has been involved in projects developing and testing sustainable food products, evaluating food waste practices among younger adults, and conducting interventions to reduce household food waste. Her research also focuses on sustainable food production in a multidisciplinary project titled 'Edible Streets', aiming to encourage urban food growing in communal spaces.

ABOUT THE CONTRIBUTORS

Ilknur Bayram currently works as Head of the Department of Foreign Languages and Turkish Education at Turkish National Police Academy. She has a PhD in Curriculum Development. Her research interests are teacher professional development, lesson study, curriculum development and evaluation, and education for sustainable development.

Sandra Bhatasara is a Scholar-Activist holding a PhD in Sociology. She is currently affiliated with Marondera University of Agricultural Sciences and Technology, Rhodes University, and African Institute for Environmental Law. She works on gender justice, social inclusion, and intersectional issues around politics, land, livelihoods, and climate change in African spaces. She has published widely on climate sustainable development-related issues, with one of her latest publications on sustainable development, women, and land issues. She also works with various non-profit organisations on sustainable development issues including gender equality, food security, poverty reduction, and climate action.

Özlem Canaran is currently teaching at Turkish National Police Academy, Department of Foreign Languages and Turkish Education. She received her PhD in the field of English Language Teaching. Her research interests are professional development of EFL teachers, English for academic purposes, and education for sustainable development.

Silvia Carnini Pulino is Associate Professor of Business Administration and Director of the Frank J. Guarini School of Business at John Cabot University. She has over 20 years' experience in international business, with a focus on finance, entrepreneurship, and technology. She is committed to innovation in education, both at John Cabot and as Chair of the Board of Teach for Italy, a non-profit organisation dedicated to bridging the educational gap in Italy.

Prosper Chopera is a Senior Lecturer with the Department of Nutrition Dietetics and Food Science at the University of Zimbabwe and Chair of the Research Group Nutrition and Dietetics. She is a holder of an MSc and PhD in Public Health Nutrition from Wageningen University and Research, the Netherlands. She also holds a postgraduate certificate in Clinical Nutrition from Katholieke University Leuven, Belgium. She has more than two decades of experience in the fields of micronutrients and health, food and nutrition security, and maternal and child nutrition as well as policy. She participates in projects assessing micronutrient deficiency as well as evaluating effectiveness of public health and nutrition interventions on improving selected nutrition outcomes as well as food product development.

Tim Doheny-Adams is a Biologist based in the Biology Department at the University of York and has previously worked on plant cell development and integrated pest management systems. He is currently focused on teaching and scholarship with a broad interest in designing interdisciplinary and inter-institutional teaching.

Ulrike Ehgartner is a Sociologist based at the School for Business and Society at the University of York, with a research interest in environmental issues, social inequality, and behaviour change. Currently working on FixOurFood, an interdisciplinary project on place-based food system transformation, her research employs participatory action research methods with practitioners in the food supply chain.

Michèle Favorite, PhD, at John Cabot University, is Executive Director of the Graduate Studies Center, Director of the Career Services Center, and Coordinator of Sustainability for Education Initiatives and also teaches Business and Communication. She has worked extensively in education, communication, and business across a range of industries in Europe, the Middle East, and Africa. She is a Sustainability Advocate and Promoter and is committed to raising eco-literacy and triggering positive change on sustainability challenges with diverse international stakeholders.

Jo Feehily is a Principal Lecturer at Oxford Brookes Business School and has driven the embedding of environmental sustainability across Undergraduate and Postgraduate Business and Marketing programmes. She has led projects that look to develop mechanisms for assessing global citizenship and modules and programmes which have sustainability at their core. Her profound belief is that higher education programmes have a critical role in developing graduates who can make a positive contribution to society and to tackling climate change, through their professional practice.

Alberto Fiore is a Professor in Food Chemistry and Technology at Abertay University in Dundee, Scotland. He has extensive knowledge in analytical chemistry, mass spectrometry, and food chemistry. He holds the prestigious title of CChem from the Royal Society of Chemistry and is a proud member of the International Maillard Reaction Society. His primary research focus lies in mitigating the formation of toxic compounds by utilising polyphenols to trap their precursors. His dedication to collaboration extends to his extensive work with African countries, particularly on projects related to nutrition and insect research. His expertise and experience in partnering with the food industry have resulted in innovative solutions to real-world problems. He has acquired research funding from prestigious sources such as European funding bodies, charities, and the UK Research and Innovation.

Sara Grafenauer is an Accredited Practising Dietitian and Fellow of Dietitians Australia. She is the Discipline Lead for the Nutrition, Dietetics, and Food Innovation Program at the University of New South Wales and was instrumental in the design of the programme. She represents her Faculty in the university's Sustainable Development Goals steering committee. Her research interests include dietary patterns, nutrition economics, consumer, and food system research with a focus on whole grain and legume foods. She has led changes to food standards in relation to dietary fibre analysis and submissions informing policy decisions. She is a Member of the Governing Board for the Global Whole Grain Initiative and is a Member of the Oat Council of Australia. Although an academic, she maintains a private practice where she consults with a range of clients, a valuable link to the practice of dietetics for student learning and teaching.

Zeng Guojun has long been committed to long-term research on food geography and food culture, and in recent years, he has devoted himself to geographical interpretation of typical and unique food culture phenomena in China. With a global perspective, he has completed a series of studies related to place and diet in line with the trend of new cultural geography. As the person in charge of the Chinese side, the applicant co-chaired the project 'Sustainable Consumption, Middle class and Food Ethics in the Global South' funded by the UK Economic and Social Council with collaborators from first-class universities in the United Kingdom, Brazil, and South Africa. He has chaired three National Natural Science Foundation projects related to food systems, and the academic team he leads has accumulated some experience in the field of dietary geography. He has published more than 60 related academic papers and published related monographs in Palgrave, the Commercial Press, and other publishing houses.

Xinwu He is a Lecturer in Accounting at Queen's University Belfast. Her research is concerned with how accounting and reporting can improve organisational transparency, enhance stakeholder accountability, facilitate long-term value creation, and contribute to the SDGs. Her research interest focuses on sustainability reporting, assurance, and accountability; integrated reporting; corporate social responsibility; and SDGs disclosures. Moreover, she aspires to integrate the research on the SDGs into the accounting curriculum, to enhance students' sustainability literacy and influence sustainable practices.

Constanza Flores Henríquez is a Doctoral Candidate in Civil Engineering at the University of Canterbury in New Zealand working in tsunami evacuation modelling. She holds a Master's Degree in Civil Engineering from Hiroshima University, Japan, and a Bachelor's Degree in Ocean Engineering from the University of Valparaíso, Chile. She has worked as a Consultant in water resources engineering and was an intern, and later Consultant, in the Tsunami unit at UNESCO's Intergovernmental Oceanographic Commission.

Matt Hopley is a fourth-year Bachelor of Science undergraduate student at Queen's University, Canada, studying Kinesiology. After volunteering with the Soul Food delivering food for two years, he is now one of two Co-Chairs who oversee all the club's operations.

Lin Jiahui is currently a Doctoral Candidate at the School of Tourism at Sun Yat-sen University. Her research is on sustainable eating practices for urban Chinese. In recent years, she has focused on the rapidly changing food environment in China and how urban residents are coping with the increasingly complex food environment in their daily food practices in order to build a nutritious and healthy sustainable diet.

Meri Juntti is an Associate Professor of Environmental Governance in the Department of Law and Social Sciences, Middlesex University, UK. Her research focuses on the role of policy and planning in sustainable rural development and urban greening.

Alexander Archippus Kalimbira is an Associate Professor of Human Nutrition and Head, Department of Human Nutrition, and Health Faculty of Food and Human Sciences at LUANAR. He currently leads and collaborates on several research projects including Geo-Nutrition Project, in collaboration with University of Nottingham, LSHTM, University of Malawi (COM), Addis Ababa University, CIMMYT, BBSRC; and Leveraging Local Capacity to Strengthen Health Service Delivery – Nutrition for Health Activity. His recent book publications was *Building the Evidence on the*

Agriculture-Nutrition Nexus: Malawi. CTA Working Paper 16/0X February 2016. ISBN 978-982-9003-88-1. Amsterdam: CTA.

William Kasapila is an Academician in the Department of Food Science and Technology at Lilongwe University of Agriculture and Natural Resources. His research interest and focus are on biofortifi-cation, fermentation, product development and innovation, sensory acceptability, food systems, food packaging and labelling, food legislation, food habits, and consumer behaviours. He has published manuscripts in the field with reputable journals. Additionally, he has published a book on *Young Adults' Dining in Hatfield: Young Adults' Satisfaction Regarding their Dining Experience in Casual Dining Restaurants in Hatfield Pretoria.*

Samson Pilanazo Katengeza holds a PhD Degree in Economics from the Norwegian University of Life Sciences, Norway, obtained in 2018. He is the Director of Research and Outreach for the Lilongwe University of Agriculture and Natural Resources (LUANAR) responsible for providing strategic leadership on the University's fundamental mission of conducting challenge-led research aimed at promoting a rich quality of life in Malawi and beyond. Samson has been working with LUANAR since 2012 and was promoted to the rank of Associate Professor in 2020. His specialties include food security and rural livelihood, climate change and resource economics, and agricultural marketing. Recent book chapters by him include *Impact of Farm Input Subsidies Vis-à-Vis Climate-Smart Technologies on Maize Productivity: A Tale of Smallholder Farmers in Malawi* published by Springer.

Sophia Lin is a dietitian with additional training in public health and epidemiology. She has worked on a range of Australian and international health projects aimed at understanding causes of and evaluating strategies that reduce the burden of chronic disease, its risk factors, and determinants of health. She is particularly interested in processes that help achieve health equity and social justice. She currently co-leads UNSW Urban Growers, a cross-disciplinary collective of university staff and students with a passion for urban agriculture and who maintains the campus food gardens. She uses these gardens as part of teaching to develop food literacy skills in dietetic students and to build connections with the community surrounding the campus through sharing of food skills.

Lesley Macheka is the Executive Director for Innovation and Industrialisation as Marondera University of Agricultural Sciences and Technology. Lesley's research interest is on food systems, food quality management, food and nutrition security, and the nexus between climate change and food and nutrition security outcomes. He is the current chairperson of the Scaling UP Nutrition Research and Academia Platform in Zimbabwe.

Faith Manditsera is a Senior Lecturer in the Department of Food Science and Technology at Chinhoyi University of Technology. Her academic background is in food science and technology. She has more than 10 years' experience in higher education teaching, supervision, and research. She did her PhD in Food Quality Design with Wageningen University and Research. She has research interests in indigenous food systems, food safety, and health food design.

Tonderayi Mathew Matsungo is the Head of the Department of Nutrition, Dietetics and Food Sciences at the University of Zimbabwe. He has 20 years' experience in higher education training and research in the areas of nutrition, dietetics, food security, and health sciences. He is the Southern Africa Representative for the African Nutrition Society. His research interests are multidisciplinary inspired by the global food system transformation agenda with specific focus on school age and adolescent nutrition, stunting, micronutrient deficiency, nutrition sensitive agriculture and biofortification, traditional food systems, nutrition transition, and non-communicable diseases. He has keen interest on SDG tracking and is an Alumni of the Africa Nutrition Leadership Programme https://africanutritionleadership.org/. He is committed to the capacity building and mentorship of young nutrition professional in Africa.

Seema Mihrshahi, Associate Professor, Macquarie University, has a background in public health nutrition and nutrition epidemiology. Her current research focuses on community-based approaches for improving healthy eating, improving food security in vulnerable groups and research into optimal infant growth and young child feeding. She has a keen interest in research translation and her work in addressing student food insecurity has led to several publications, small grants, and changes to student welfare and education policies. She has authored over 130 peer-reviewed research articles and three book chapters relating to human rights and social policies for women and children and nutrition throughout the life course. She has participated in various research projects with a focus on women's and children's health both within Australia and internationally.

Tema Milstein is Professor of Environment & Society at University of New South Wales in Sydney, Australia. She is an internationally recognised leader in environmental communication, a transdisciplinary field that understands communication as having far-reaching effects at a time of human-generated environmental crises. She is particularly known for cultural approaches to studying how communication shapes ecological understandings, identities, and actions. Her work tends to discourses that otherwise go unnoticed, to connections between discourses and wider destructive or restorative practices, and to existing and emerging paths towards sustainable, just, and regenerative futures. Her co-edited books include the Routledge Handbook of Ecocultural Identity (2020) and Environmental Communication Pedagogy and Practice (2017). She is a Former Fulbright Scholar and Recipient of the Faculty of Arts, Design, and Architecture Dean's Research Award for Society Impact.

Vincent Mlotha holds a PhD in Aquaculture from Stellenbosch University in South Africa. His research interests are on nutrient composition and contribution of fish to food security and nutrition in developing countries. He is a Research Scientist for food composition data compilation and management for the Malawi Food Data Systems (MaFoods) programme and serves as the Deputy Head for the Department of Food Science and Technology at LUANAR.

Xenia K. Morin, PhD is an Associate Teaching Professor in the Plant Biology Department at Rutgers University – New Brunswick. She teaches courses in the agriculture and food systems programme.

Samuel Mwango holds a Master of Science in Nutrition from the Centre of Excellence for Nutrition at the North-West University South Africa. He is a Keen Researcher in Clinical Nutrition and has two publications.

Agnes Mbachi Mwangwela is an Associate Professor of Food Science with 28 years of experience in university teaching in food science, food product development, and food analysis at Lilongwe University of Agriculture and Natural Resources. She is a Holder of PhD in Food Science from the University of Pretoria, an MSc and BSc from the University of Malawi. She currently leads food composition data compilation and management for Malawi Food Data Systems (MaFoods) and is a seasoned researcher who has published and has made presentations of her work in international conferences. Her research work is on the chemistry, utilisation, and safety of legumes and cereals.

Karen Oberer is a Sustainability Officer (Monitoring and Reporting) at McGill University in Montreal. She is responsible for collecting data and reporting on McGill's sustainability advancements by working with partners from across the university campuses. In this role, she has tracked the university's contributions to the SDGs, which were featured in Canada's 2021 Annual Report on the 2030 Agenda and the Sustainable Development Goals. She has a PhD in English Literature from McGill University and an MA in Environmental Practice from Royal Roads University.

Juan Diego Zamudio Padilla is a Doctoral Candidate in Economics at Hiroshima University, specialising in Experimental Economics. He earned a Master's Degree in economics at Hiroshima University. He is a Professor of Economics and Founder and Researcher at the San Marcos Centre for Asian Studies, in the first macro public university in Peru, Universidad Nacional Mayor de San Marcos, with 7 years of expertise in international higher education and

10 years in economic analysis. He serves as an Advisor in financial matters and the internationalisation of higher education.

Jessica Kampanje Phiri is a Social Anthropologist specialising in understanding the socio-cultural dimension of poverty and food systems from an inclusive and community engagement perspective. Currently employed as a Senior Lecturer at the Lilongwe University of Agriculture and Natural Resources, she brings with her more than 10 years' experience in interdisciplinary and multi-country-related research. Her recent work includes being the Malawi Lead Researcher on a Multi-Million Dollar Climate Change-related project in collaboration with University of Cape Town; A Co-PI on a Feed the Future's Innovation Lab for Crop Improvement Center for East and Southern Africa in collaboration with Cornell University; and a Social Analyst for the Prioritizing Options for Women's Empowerment and Resilience in Food Tree Value Chains in Malawi Project in collaboration with ICRAF, GLOW-CDKN, and ODI.

Viren Ranawana is a Senior Lecturer in Human Nutrition at the University of Sheffield. Having a background in food science and technology and in nutrition, his research focuses on the food–nutrition interface. He has a particular interest in developing sustainable diet-based strategies for mitigating chronic diseases and malnutrition. He also has an interest in the use of underutilised foods for improving food security, health, and nutrition status.

Thomas Rosen-Molina provides high-level programmatic and analytical support to Office of Global Affairs' initiatives in the University of California, Davis, United States. Prior to this position, he served as a Foreign Service Officer with the US Department of State, where his roles included Consular Officer, Human Rights Officer, and Refugee Coordinator in Montenegro and Political Officer for External and Multilateral Affairs in Germany.

Cathy Sherry is a Professor in Macquarie Law School and Executive Member of Smart Green Cities and the Centre for Environmental Law. She is a leading international expert in land law, with a particular focus on high-density development. She advises governments around the world on their apartment law. She has a particular interest in liveable cities, and the importance of green infrastructure, including urban agriculture. She was the Team Lead of UNSW Urban Growers, a cross disciplinary group that creates food growing spaces on a high-density campus. She convened and co-taught Food Law, a course that focuses on problems in modern food systems. She taught students to grow their own food and to increase their food literacy and their understanding of the precarity of food production. She has published research on the role of university food gardens in augmenting graduate environmental literacy.

Jolynn Shoemaker is the Director of Global Engagements in Global Affairs at University of California, Davis, United States. In this role, she develops strategy and programmes to advance the UN sustainable development goals and other international initiatives. In addition, she is a Fellow at Our Secure Future: Women Make the Difference, where she focuses on women, peace, and security. She spent two decades in Washington, DC, working on foreign policy and national security issues. Her prior experience in federal government includes both the US Department of Defense, Office of the Secretary of Defense, and the U.S. Department of State, Bureau of Democracy, Human Rights, and Labor. She has worked with a variety of non-profit organisations engaged in policy-relevant research, advocacy, and training.

James Stockdale is a Biogeochemist at the University of York with a background in greenhouse gas flux research in agri-ecosystems and carbon-rich natural systems. After working as Knowledge Exchange Fellow on a large interdisciplinary programme, N8Agri-Food, he has since co-founded and leads a spin-out company aiming to design and supply environmental monitoring technologies to a wider range of users. He combines this role with teaching in the Department of Biology with a focus on increasing students and engagement with the translation and impact of research, including food system-based and bio-entrepreneurial modules.

Quyen Vu Thi is the Head of Biotechnology Department in the Faculty of Applied Technology, Van Lang University, Vietnam. Her research focuses on the application of biotechnology in agriculture, food processing, and pharmaceutical raw materials.

.

www.ingramcontent.com/pod-product-compliance
Lightning Source LLC
Chambersburg PA
CBHW071740270326
41928CB00013B/2748